What Nature Does for Britain

First published in the UK in January 2015 by

Profile Books

3A Exmouth House
Pine Street, Exmouth Market
London, EC1R OJH

Typeset in Bembo, THE Sans and Crete Round to a design by Henry Iles.

10 9 8 7 6 5 4 3 2 1

Printed in the UK by CPI Group (UK) Ltd, Croydon, CR0 4YY
on Forest Stewardship Council (mixed sources) certified paper.

A catalogue record for this book is available from the British Library.

ISBN 978 178125 3281
eISBN 978 178283 0986

What Nature Does for Britain

Tony Juniper

ADDITIONAL RESEARCH BY
Lucy McRobert

PROFILE BOOKS

The Rutland ospreys – a national asset.

Contents

Preface

What Nature Does For Britain

'Money is no object in this relief effort.
Whatever money is needed, we will spend it.'

Those were the Prime Minister's words: a promise of open-ended spending, at the height of economic austerity, in response to the widespread flood damage of 2014. It was not surprising, of course, given the terrible circumstances facing thousands of people across the country, whose homes and businesses had been inundated with stinking muddy water. But research suggests that reacting to flooding in this way – throwing money at cleaning up the mess after the event – is the least rational and most expensive way to go. It would be better by far to see the bigger picture – how the changing climate is likely to cause more such events in future – and prepare for that in far more intelligent ways, including rebuilding what might be called 'green infrastructure'. That is, restoring the natural systems that provide so many valuable services, including some level of protection from flooding.

Yet despite a mounting body of evidence to show how we can save money, protect property, promote wellbeing and create more value through looking after nature, there is a widespread view that doing so is somehow hostile to Britain's interests. You hear it from politicians, economists and columnists, and most notoriously from David Cameron himself, who is said to have told his aides to 'get rid of all the green crap' – only four years after promising to lead 'the greenest government ever'.

The fact that so many people seem to accept this line of thinking – that the protection of nature is harmful to people and the economy – is the reason for this book. On the eve of a general election in the UK, I will set out to show that rather than being a barrier to progress, Britain's nature is an economic and security asset with enormous social value. At the end of each chapter I will offer proposals that I believe would help the recovery of nature in Britain and in so doing help the economy and people. I have put these in the form of 'manifesto' points and I offer them freely to all of our politicians to adopt. My hope is that these ideas will inspire some of our elected representatives to see what can be done in making sure Britain's nature retains its value in the decades ahead.

In the pages that follow I'll take you on a journey around Britain. But not through the familiar political landscapes. We are going instead to the real Britain, the one where we are supported by nature, wildlife and natural systems at almost every turn. We'll look at the country's flood defences, for example. Not the costly man-made ones but the ones that nature endowed us with, and which can be protected or revived for modest sums: soils and wetlands, renaturalised rivers and coastal marshes among them. These and other natural assets, properly maintained, can hold more water in the environment and reduce the risk of floods.

We'll visit places where the protection of natural habitats can also provide a cleaner, cheaper water supply. Bogs, woods and marshes strip out sediments and pollution, meaning that less high-tech equipment is needed to clean it up before being piped to the public. From the wild blanket bogs of Northern Ireland to the arable farmlands of southern England we'll see how looking after nature can support our water security, and in highly cost-effective ways.

We'll take a look beneath our feet, too, at one of the least appreciated aspects of our natural heritage, namely healthy soils. These help to purify water and reduce flooding and are a massive store of carbon, which if kept intact and enhanced can make a big contribution to Britain's efforts to combat climate change. Woodlands, wetlands and dune habitats are similarly locking up carbon and keeping it out of the air, at the same time as providing a range of other important services.

Those same soils that hold carbon and water are also, of course, the source of most of our food, supporting jobs and presenting a wide range of economic opportunities. We'll see why it's rational to keep those soils in good health, and how that in turn improves the quality of our rivers and seas, enhancing recreation and tourism. We will observe how our food security is also enhanced by wildlife, including the hundreds of species of native bees that live in Britain and the host of spiders, beetles, bats, birds and other predators that help to control pests.

The oceans that surround our country are also a source of food, jobs and economic opportunity, not least through the billions of pounds-worth of seafood we take each year. They are under threat from poor political management, but not without hope, if among other things we scale up the positive examples of fishermen boosting their incomes by using better and more sustainable fishing methods.

We'll see how despite becoming ever more used to seeing our nutritional security through long-distance global food-supply chains, for most of our food we remain fundamentally dependent on nature here in the UK – whether it's the plankton-fuelled food chains that supply seafood, the intricate ecological relationships that enable soils to recycle nutrients, or the 140-million-year-long partnership between insects and flowering plants and nature's predator and prey struggles. Wildlife remains essential, even for the majority of us who live in Britain's cities, in providing much of what we eat. And the foods we get from the sea and with the help of pollinators – fruit, vegetables and fish – include the healthiest dietary choices we can make.

Our seas also present vast opportunities for increasing our energy security. The wave, tidal and wind energy resources available in our marine environment present job-creation prospects while at the same time helping reduce the risks to our economy that come from dependence on energy resources from unstable regions. On land we can use our woodlands, the wind, sun and waste to secure supplies of sustainable energy, driving technological innovation and creating employment in the process.

By supplying a higher proportion of our energy from clean and renewable sources we'll reduce our carbon emissions – as we are bound to do by the UK's own Climate Change Act – and in so doing add to the fundamental wellbeing of the entire world. But this need not be a burden. In addition to reducing emissions and creating jobs, clean energy will also result in less health-threatening pollution, easing demands on the National Health Service. This is not the only way working with nature can promote wellbeing. We'll also see how increased access to good-quality natural areas near where people live helps reduce mental illness and can even be a factor that cuts crime.

In the Britain we're about to visit, businesses and citizens not only rely on nature but have huge potential to flourish through a more thoughtful partnership with it. Our water companies, the NHS, the small and medium-sized companies based on farming and fishing, supermarkets, the businesses that supply an increasing proportion of power from renewable sources, insurance companies, drinks manufacturers and even those making cars are among those who depend in different ways on healthy nature to make profits and sustain employment. In short, we'll see how a thriving UK plc depends fundamentally on the health of Nature Ltd (UK).

There is no doubt that nature's practical value is huge, though assigning precise economic values is complex. The Office for National Statistics made a first attempt to put a monetary values on 'natural capital' for the UK in 2011 and estimated these at £1,573 billion (that is over £1.5 trillion). That number is based on experimental methods and is open to critical review, but it surely reveals the scale of falsehood at the heart of the political debate, where we continue to be presented with an apparent choice between looking after nature on the one hand or growing our economy on the other.

And to return to David Cameron's pledge, the good news is that money really is no object: we just need to harness the resources we already have – in taxes, bills, subsidies and investments – more intelligently towards longer-term, sustainable aims.

Tony Juniper, January 2015

SOURCES

This book draws on my own 30 years of experience in making a case for nature, as well as a large body of recent science and the views of many different experts. In order to create a readable narrative, I have not included footnotes or references in the book but have instead compiled a compendium of source material at my website. If you want to find out more, you can go there and, for the most part, click directly through to original sources. **www.tonyjuniper.com**

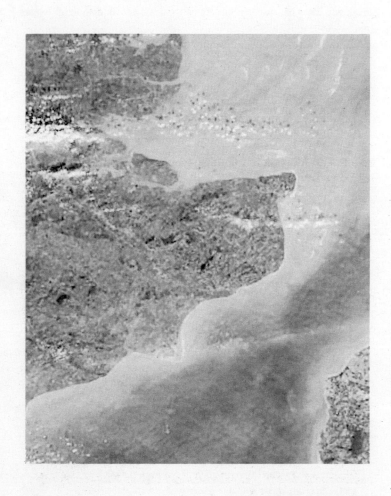

This satellite image, taken in February 2014, shows where our soil goes once it's washed off our fields.

Chapter 1
Nutrient Nation

BETWEEN £900 MILLION AND £1.4 BILLION – THE ANNUAL COST TO THE UK
ARISING FROM SOIL DEGRADATION

THREE QUARTERS – THE PROPORTION OF TEMPERATE FOODS WE CONSUME
THAT ARE PRODUCED ON BRITISH SOILS

12 MILLION TONNES – THE QUANTITY OF ORGANIC MATERIAL DISPOSED OF
IN LANDFILL EACH YEAR IN BRITAIN

March 2014. British TV viewers are treated to amazing
pictures of southern Britain taken from the International Space
Station. Orbiting more than 330 kilometres above the point
where the Atlantic meets the foaming rocky coast and green
fields of Cornwall, the film sequence begins over Land's End.
Travelling at around 27,000 kilometres per hour the spaceship
takes less than two minutes to traverse the whole south of
England before heading out over the North Sea. As its orbit
takes the camera over London and the Thames Estuary a very
noticeable broad brown fringe of sea can be seen hugging the
east coast. 'Run-off', remarks astronaut Mike Massimino as he
looks out through a window. His observation is almost an aside.
It is for him evidently a familiar sight.

Some miles out from the shore and the water is blue,
but where sea meets land, and particularly where the rivers
discharge, the muddy colour is pervasive. The rivers running
into the ocean along the east coast, including the Thames,

mostly drain farmland. Much of it is subject to intensive cultivation for arable crops. The repeated ploughing disturbs the soil, and that is what that 'run-off' is – soil that has left the fields and is now in the sea.

Soil erosion is one of those phenomena that rarely make headlines, though it is profound in its implications. Stories linked with it often do, however, even if most of us don't notice the connection. One is flooding, another is the price of water. And a third is the cost of food. It is easy to forget that one of the most important functions provided by nature to our economy is the recycling of nutrients. In fact, so vital is the flow and replenishment of nutrients that most of our country looks the way it does today because of our attempts over millennia to maximise the benefits of how nature does this. Key to the whole thing is perhaps our least appreciated natural asset – soil.

Forests, farming and fertility

The little town of Dyffren Ardudwy lies behind a strip of sand dunes on the Irish Sea coast of Gwynedd in northwest Wales. Climbing out of town on the Fford y Briwg the land rises steeply to reveal spectacular views of the sea, the Lleyn Peninsula and mountains of Snowdonia.

About three kilometres from the sea, and at about 150 metres above the long sandy beach below, the land levels out. On this undulating but flatter part of the hillside there is a patchwork of fields separated by thick stonewalls. While landscapes superficially similar to this are not unusual across the British Isles, especially in some of the uplands, this place is special.

Constructed from stones cleared from the land during preparations for agriculture, some of the walls are six feet thick and founded on huge boulders. Their origin is thought to be in the late Stone Age – the Neolithic – and they are therefore some of the oldest relics of settled agriculture in these islands. Piles of fire-cracked rocks believed to have been used in communal cooking or bathing, probably in the Bronze Age (around 4,600 years old), remain in the fields.

It's windswept up on the hillside and feels exposed. Stunted oaks have been sculpted by steady westerly winds and periodic gales to lean landward. A trio of Welsh black cattle graze in a damp hawthorn-clad hollow while groups of sheep totter ahead of me around the Bronze Age mounds. The temperate aspect and sheets of grey mist that hang off the slopes above reveal two of the conditions necessary for agriculture – namely mild conditions and water. A third is seen in the weak midwinter sun. Close to the horizon and casting sharp long shadows along the walls and behind the mounds of stones, our home star has just passed its nadir and is now returning earlier each morning and setting later each evening. In the months ahead it will increasingly power plant growth and once more warm that other vital prerequisite for farming: the soil.

Before fields emerged between the lines of stones laid out by Neolithic and Bronze Age farmers, this part of Wales, in common with nearly all of the rest of the British Isles, was covered with trees. A little to the south around Barmouth some of the hillsides are still clothed with extensive patches of sessile oak woods, providing some suggestion of what this entire landscape would have looked like before the native rainforests were cleared to make way for farming.

The fundamental change from hunting and gathering and toward farming happened at different times in different places. Originating in Syria and Iraq between about 13,000 and 11,000 years ago, agriculture spread out across Europe to reach Britain between about 7,000 and 6,500 years ago.

It took about 2,000 years more for farming to spread right across the British Isles, but once it got going there was no turning back.

The trees were killed through cutting away a ring of bark around their trunks, felling with axes and by burning. Traces of the forest fires set by these first British farmers are evident today in faint layers of charcoal that can be detected in some soils. Other evidence pointing to great change is seen in pollen archives. When flowers shed their microscopic genetic capsules some fall on lakes and bogs where they become preserved in sediments. Each plant species' pollen is distinctive and today we can read the tiny grains like an ecological diary that sets out in chronological order the changes that occurred in the vegetation that covered our islands over thousands of summers past. The pollen record tells the story of how there was a dramatic reduction in tree cover and the rise of more open environments – ones created by people for producing food.

The availability of land was initially not a major limiting factor for farming. It was the ability to clear it that presented more problems. During the late Stone Age agriculture more akin to 'slash and burn' than to modern cultivation prevailed. Ploughing and planting between burnt and ring-barked trees, farmers moved on once the soil became exhausted, returning to plant again when it had naturally recovered fertility. Over time, however, population pressures meant that farming the same land again and again over long periods became necessary. No longer could people rely on moving from one plot to another, mining the fertility as they went; they had to invest effort in replenishing the soils that had been claimed from beneath the forests.

Soil is a little word that describes a complex thing. Its character in any one place arises from the interaction between many factors, including the fundamentals of the geology upon which it rests. Another determinant is climate. Wet and cool conditions lead to different soils compared with where it is warmer and drier, even when the rocks beneath are the same.

The shape of the ground is important too, with steeper slopes tending to have thinner soils than valley bottoms. Then there is the time the soil has had to form. The longer it has been building up in any one place, the deeper it becomes. The character of the organisms living in it, on it and from it, and the amount of once living, but now dead, material the soil contains – the organic matter – also influence what it is like. Then there is a sixth very important factor: what we have done to it.

For millennia, people have been clearing and changing natural habitats across the British Isles so as to make use of the soils beneath them, and today farmland enclosed in fields takes up more than half of the United Kingdom's total land area. Forests, upland bogs, unenclosed pastures, urban areas and areas set aside for nature conservation account for much of the rest.

Thousands of years after the rise of farming, most of the UK's soils are still harnessed for the production of food, in the process contributing about two billion pounds toward gross domestic product and directly supporting about half a million jobs in farming. But of most importance to us all is the fact that we produce about 60 per cent of our total food needs from these soils, and about three quarters of the temperate produce we consume.

Whatever the soil type, clearing natural habitats and then ploughing and grazing the ground beneath cause dramatic changes, especially in relation to soil nutrients. Nutrients are the substances plants need to make the cellular machinery that enables them to capture and convert sunlight into complex chemicals, and thereby to grow and reproduce. They include nitrogen, phosphorus and potassium (often referred to as NPK), although more than a dozen others are also essential, including carbon (derived from carbon dioxide in the air) and hydrogen (from water).

In a natural habitat, such as a forest, the accumulation of soil nutrients occurs over hundreds or thousands of years. Held in plants and animals, as well as the soil, nutrients largely stay there, cycling through different parts of the system. It is more or less a closed loop, the essential substances remaining in the forest. When a tree dies, its trunk, branches and leaves are broken down by fungi and other organisms into their constituent parts. Through this process of decay, the nutrients it took to grow the plants in the first place are released back into the system, and through the soil returned to living organisms.

There is a little fragment of ancient woodland near to where I live in Cambridgeshire called Hayley Wood. A visit there in autumn can induce a rather melancholy mood. But for the sad-sounding song of a robin proclaiming its winter territory, all is quiet, and in the cool gloom of early evening a steady stream of golden leaves falls. It seems death and decay has seized this once vibrant world. This time of waning life is, however, a vital step in a bigger cycle, one that is as important as all the others, for without this period of brown, red and gold, there can be no green. A return trip six months later, in early May, reveals pale fluffy leaves bursting from the buds on the oak and ash trees. A blue haze infuses the otherwise emerald ground layer. Leaves that fell last October and November are in the soil, dragged down by earthworms, where a recycling process is, with the help of a multitude of microbes, well under way.

The miraculous work of bacteria, microscopic fungi and others, and the once living but now decaying remains of dead plants, renews the life cycle. As the breakdown of organic material refuels the soil engine, so the burst of bluebells above marks the start of summer. With each day the light lasts longer and we are reminded how the whole intricate system of nutrient cycling is powered by sunshine.

In an old wood like this one the dense roots of the trees and other plants form a tight net that is among other things a very effective nutrient recovery system. The microscopic fibres and hairs on the roots catch nutrients released by decay. So tight and efficient is nutrient recapture that even during heavy rain few of the essential substances are washed out and escape into the wider environment. The system evolved to sustain itself by ensuring there is no waste. The recycling rate is near to 100 per cent.

When it comes to what were once woodland soils that are now cultivated for agriculture, a very different situation prevails. Instead of closed loops, most farms today are linear systems. Nutrients are added, taken out again in crops (or lost), and new ones must be put in from outside, otherwise food production would come to an abrupt end.

This basic fact of life has of course been appreciated for some time. As they laid out their fields with the hard edges of human organisation imposed upon the wavy lines of nature, Bronze Age farmers realised that they had to do more than put boundaries around parcels of land to produce food. They also needed to find new nutrients to replace those removed in crops and livestock. Their answer was manure. Nutrients captured in rotting plants or those salvaged from the far end of animals' digestive tracts were among the sources that replenished depleted fertility, and thus enabled more food to be grown.

For thousands of years manure and bones (especially rich in phosphate) sustained food output. But when in the nineteenth century an agricultural revolution swept across Britain, with mechanised planting and harvesting powered increasingly by steam-driven machinery, new rotation methods and more effective selective breeding, demand for nutrients rocketed.

Phosphate supplies from Britain's bone-breaking knacker's yards couldn't keep pace and other sources were sought. Indicative of how desperate the need for this nutrient became was the import of tens of thousands of mummified cats from Egypt's ancient tombs and pyramids. The sun-bleached bones of animals that had died in the North African desert were collected and shipped to Britain too. Even the battle-fields of Waterloo and The Crimea were scoured for skeletal remains. All of this was ground up and processed into fertiliser. One commentator remarked at the time how 'Great Britain is like a ghoul, searching the continents for bones to feed its agriculture'.

Another phosphate source was oceanic islands, where guano – droppings left by vast colonies of seabirds – had over thousands of years led to the buildup of rich nutrient deposits. Imported to Britain from the Caribbean and Pacific, these concentrated phosphates bolstered food production. Then during the mid-nineteenth century phosphate sources much closer to home also came into widespread use.

This came from nodules that had accumulated at the edge of a warm Cretaceous sea, a hundred million years ago. Comprised of among other things the bones and droppings (so-called coprolites) of dinosaurs, the nodules were deposited across a broad band of eastern England. The concentration of nutrient in the nodules was very rich and the product made from them became known as 'superphosphate'. One contemporary writer dubbed the pits from where the nodules were excavated as 'food mines'. Another reason this product, made in part from dinosaur poo, was 'super' was because it was cheap. Coming from a local source rather than the other side of the world it cost only half as much as guano brought on ships from thousands of miles away.

So prolific was the industry that exports from Britain were shipped as far afield as Russia and Australia. The phosphate-mining rush enabled fortunes to be made. But as market conditions changed so the party came to an abrupt end. The industry peaked in 1876 with its output at over a quarter of a million tonnes, and then it went into rapid decline, in part due to more open trade policies that allowed Britain to import cheaper foreign grain.

But as demand for food continued to grow, and as knowledge of chemistry and plant biology improved, so new strategies for fertiliser production were adopted, especially for the NPK mix of nitrogen, phosphate and potassium.

Magic mix

During the early twentieth century the search was on for strategically important supplies of the K part – potassium. Historically, wood ash had been the main source, but as industrial-scale farming spread so bigger flows were needed. Deposits of potash left behind when saltwater bodies had evaporated became especially important. Among these were the brine-encrusted lakes of the Nebraska Sand Hills, deposits found by the Dead Sea and substantial finds in Canada.

At about the same time as these and other sources of potassium were being fed into fertiliser factories a process was invented for making industrial quantities of nitrogen using the otherwise inert gaseous form that comprises most of the Earth's atmosphere. Nitrogen may be the most abundant gas in our air, but until 'fixed' it's useless to plants. Nature 'fixes' some nitrogen via the natural action of bacteria and through lightning strikes. But when we found a way to augment these natural sources and to fix nitrogen in fertiliser factories, and on a grand scale, everything changed.

The process enabled the creation of ammonium nitrate from natural gas (from which hydrogen is extracted to be combined with nitrogen from the air) and its large-scale adoption is perhaps one of the most important developments in human history. When ammonium nitrate was first used shortly before the First World War the global population was about one-and-a-half billion people. The fact that it has rapidly increased to over seven billion today is in no small part down to the availability of this abundant fertiliser. Such is the scale of its contribution that about half the protein within our bodies is made from nitrogen originally fixed by this industrial process.

Its use in farming became especially widespread after the Second World War, when explosives factories that had been making ammonium nitrate for bombs and shells were turned over to more peaceful purposes. British farmers were among the first to embrace a new agricultural revolution, based not only on new fertilisers but expanded cultivation, through opening up and ploughing previously wild areas. New seed varieties were introduced and so were new chemical pesticides. In its own narrow terms it worked.

Aided by different kinds of policy and financial support, the production of some food crops went up dramatically. In 1940 the British wheat yield was about two-and-a-half tonnes per hectare: today it is more typically eight. During the 1930s self-sufficiency in foods that can be grown here was roughly 40 per cent. In the post-war years it went up fast and today is at about 73 per cent. Considering the rise of the UK population, this is a doubly impressive achievement.

Fertiliser costs – nitrous oxide

The increase in food output has not come without considerable costs, however. One is seen in the high levels of climate-changing gases released by modern farming. Carbon dioxide, methane and nitrous oxide come from, among other sources, animal slurry, fertiliser and soils. Nitrous oxide from fertilisers is an especially potent climate-changing agent – molecule for molecule about two hundred times more powerful than carbon dioxide. When it drifts high up into the stratosphere this pollution also causes damage to the Earth's protective ozone shield.

At the same time nitrogen gases released from fertilisers contribute to an increased concentration of ozone in the *lower* atmosphere, where it is a pollutant rather than protector. When it gets into the air we breathe, nitrous oxide pollution can also lead to serious respiratory illness, cancer, and cardiac disease. Yet another pollution problem, acid rain, is fuelled in part by nitrogen oxides from fertiliser.

Then there is the impact that fertilisers cause when they escape from fields to enrich the natural environment. Some of the nitrous oxide that goes into the air falls down to fertilise not crops but different kinds of natural habitat, such as grasslands. This can reduce diversity, as more aggressive common species make better use of the nutrient windfall, in the process wiping out more fragile ones.

Nitrogen and phosphate are also carried away in water to affect aquatic and marine ecosystems. Unlike the fine-meshed root net found in forest soils, where there is a close association between plant roots and the rich diversity of microorganisms living in the ground, intensively farmed soils are highly simplified and some of the fertiliser goes right past the roots and departs the fields.

Humans have more than doubled the amount of biologically active nitrogen in the world today, marking a bigger change than even the amount of carbon dioxide we have added to the atmosphere. As the different impacts from releasing so much manufactured nitrogen have become clearer, so there have been some attempts in the UK to reduce these, with significant progress made both in the amount used and how it is applied. The scale of nitrogen loss is, however, still huge.

The rapid rise in the human population has also been made possible through more and more phosphate being made available. Phosphate use in fertiliser also causes environmental impacts but with this vital nutrient come additional challenges arising from the fact that the geological deposits we rely on are being rapidly depleted, and unlike nitrogen fertiliser, we can't make an alternative supply from thin air. Some analysts estimate that phosphate output might no longer be able to keep pace with rising demand as early as the 2030s, bringing serious implications for food security.

As well as questions linked with how we can manage the impacts of nutrients escaping from the food system into the wider environment, and how to sustain supplies of finite resources that are being rapidly used up, there is the matter of wider soil health, and how this vital prerequisite for plant growth can be kept in good shape.

Cultivation, cropping, the use of industrially produced fertiliser and use of pesticides have altered the nature of many farmed soils. A reduction in the proportion of organic matter is one clear trend. As natural cycles of nutrient replenishment are replaced with the linear application of manufactured inputs, soils become less complex and more a medium in which plants can simply be fed nutrients from outside.

According to Stephen Nortcliff, Emeritus Professor of Soil Science at the University of Reading, this is a concern in many parts of the UK's agricultural land, in particular in the cereal-producing areas, where he says there are 'relatively low levels of soil organic matter and the possibility that these levels are declining because insufficient organic material is being returned to the soil. The organic matter has a number of influences on the behaviour of soils, such as the ability to retain water and the development of soil structure. Structure is important because it has a major influence on the infiltration of water into the soil and seed emergence, among other things. The decline in soil structure may result in soil erosion which is environmentally very damaging and expensive to farmers and local authorities.'

While Nortcliff's observations relate to changes taking place during the recent past, it is perhaps not surprising to know, considering how long farming has been going on across Britain, that soil damage has been happening for a very long time too. It started with woodland clearance in the Neolithic. On the South Downs for example the covering of light wind-blown silt that had built up there at the end of the last Ice Age was lost as a result of farming during the Bronze and Iron Ages, giving rise to the stony soils of less than 20 centimetres' depth that are found there today. In many other areas too historic soil damage is evident. It continues today, although now on a con-siderably grander scale, and is costing us a lot of money.

Losing a billion a year

Professor Jane Rickson at Cranfield University, another of Britain's leading soil scientists, has been involved with work looking at the overall economics of soil health.

'We degrade our soils by the way we use them. We use them beyond their capability, beyond their capacity,' she says. Just how much we are doing that is hard to answer, as ongoing survey work has been neglected in recent decades. But Rickson notes that 'we're observing with the changing weather patterns, such as more intense rainfall, increased soil erosion. This is both in the magnitude of that erosion and also in the extent, so that fields that have never eroded before are now doing so.'

This is not only down to more extreme weather but also changed farming practices, including the trend for the planting of more winter cereals. These crops cause soils to be left bare during the autumn, often the wettest time of year.

Rickson points out how we have managed to mask the effects of erosion and diminished soil health with more fertiliser and, to an extent, irrigation. She says that while many farmers are aware of this, because of economic and market conditions they have to carry on with practices that cause soil loss and degradation. She says that there are a number of steps that can be taken, if only farmers have the means. 'You can use more compost; you can put more drainage lines in that would help reduce the amount of surface runoff; you could put buffer strips in; you could try and increase your organic matter.'

This begs the question of how farmers can be incentivised and encouraged to do these things, especially those struggling to make profits working marginal land. We'll come back to that later. Meantime it is instructive to understand the estimated costs that come with how we treat soils.

Despite the patchy nature of information on soil health in the UK today, Cranfield scientists have been able to estimate that for those impacts that can be quantified in monetary terms the costs of soil degradation to the UK are between £900 million and £1.4 billion per year. Nearly half of these costs are caused by the loss of organic matter, over a third by compaction and about 13 per cent due to erosion. In terms of the services that we benefit from and which are being degraded in the process, food production is one, but as we'll see later, big costs are also being felt in the form of increased greenhouse gas emissions, increased flood risk and reduced water quality.

While this context should give us more and more reason to be concerned about the deteriorating state of soils across much of the UK (and indeed the world), it seems most of us have yet to appreciate the profound implications of this broad trend.

Stephen Nortcliff has thought about that a lot and reached the conclusion that soil is central to our entire social and economic system. 'Just like you have a keystone in an arch that holds these enormous structures together with amazing rigidity. Take that one stone away and the whole thing collapses. My contention is that the soil in many respects works within the environment like an architectural keystone. Its presence keeps the system functioning, and when we remove it or make it less well-fitting, things start to go wrong. If we don't look after the soil I would argue that the system would become unsustainable and collapse.'

Most policy-makers don't see this at all, although one who does is Baroness Sue Miller. She is a member of the Parliamentary Agroecology Group which brings MPs and Lords together in considering the sustainability of agriculture, including in relation to soils. 'It takes between a thousand and ten thousand years to form a significant layer of topsoil,' she asserts. 'Yet we can destroy it in a couple of generations and that's it.' Given the fundamental importance of soils she believes they should be regarded as a strategic national asset, 'not just something that people grow their crops in today'. She calls for a stronger research focus on soil health, investigations into the effects of different agricultural chemicals and more work on how best to restore damaged soils.

One person who agrees with all that is farmer Iain Tolhurst, Director of the Tolhurst Organic Partnership.

A farm that mimics woodland

From a gentle south-facing slope overlooking the River Thames I can see the silvery ribbon of the great river sparkle about a half-mile distant. Shafts of wintry sunshine strike down through holes in the low clouds like golden searchlights, brightening patches of the broad floodplain. The levels that flank the river's channel, smoothed by meanders over thousands of years, tell of how in past times the Thames was much closer to where I am standing. Just behind the field to the north is a reminder of why it was also once much bigger.

There rises the steep scarp slope that marks the edge of the Chiltern Hills. Today this part of South Oxfordshire is covered with rich chalk grasslands and lush beech woods, but up there 10,000 years ago was the southern edge of the ice sheet that covered most of Britain. When it melted the Thames swelled into a mighty raging torrent that flowed east to become a tributary of what is today the River Rhine. It is upon the gravelly deposits left behind by these post-glacial upheavals that Iain Tolhurst farms.

Derived from chalk and flinty debris left behind by ice and river, the thin soils have never been ideal for growing crops, and they were degraded further by agriculture. That is, until 1987, when Tolhurst arrived and began a programme of soil restoration. On a bright December morning, he explained to me what he was doing:. 'We produce about a hundred different kinds of vegetables during the course of the year. That is about as diverse as you can possibly manage without the use of heated greenhouses. Everything we do is about soil health and everything we do in our business has to fit in around that.'

Several methods have helped the soil to recover, including the use of green manures and carefully planned crop rotations that capture and recycle nutrients. The reason Tolhurst has gone down this route, rather than importing large amounts of manufactured fertilisers, is to achieve a durable and resilient farming system. By that, he means farming that can continue indefinitely, and which deals with the different impacts it could otherwise cause. Keeping nutrients in the system is key to his plan. 'Making this farm work has required us to plug all the gaps from where nutrients were escaping. We're trying to keep the fertility in the system. One way we do that is to make sure there is no bare soil. That's why we plant green manure, such as clover, grasses or different weeds, to capture nutrients and stop them escaping.'

While for most growers the idea of planning weeds would be anathema, this approach has enabled Tolhurst to avoid using the kind of industrial quantities of fertiliser that keep most of the larger farms in his part of the country in business, including the ammonium nitrate that is used to boost nitrogen levels. He explains that with these different methods farming can actually get away with using fewer added nutrients. 'In low-input systems plant roots grow in a more natural and healthy way and they find what they need. But if you feed plants lots of nutrients from above, their roots will stay at the top of the soil and they won't go looking for more. So you can get away with lower nutrients, but only if the soil is healthy.'

I ask him what he regards as the best indicators of soil health. He has two: 'one is microbiological activity and the other is the carbon level.' Both of these are components of organic matter. Over time, and through careful management, the soil organic matter in his fields has steadily increased, bucking a nationwide trend. 'It can hold moisture when it's dry and allow drainage when it's wet, it also holds soil microbes, which is particularly important, and also helps to hold nutrients.'

One indicator of rising levels of organic matter, and therefore soil health, is earthworm activity. We walk across areas planted with strawberries and leeks where the ground is almost completely covered with wormcasts. The crumbly dark brown squiggly little mounds are comprised of material that has passed through these remarkable creatures – the primary processors of organic recycling, turning dead leaves and other plant material back into soil. Tolhurst explains how there are thousands of worms per cubic metre and the silty substance that was ground up in their guts is richer in nutrients than the surrounding soil and is being spread over the top by the worms. Whereas most modern farming is about feeding plants, the approach here is based on feeding the soil, and that in turn means feeding the worms. Tolhurst takes up a worm cast and crumbles it in his hand. He can barely contain his excitement as he tells me of the work the worms are doing in sustaining the health of the land.

The worms are doing that by, among other things, eating the green manure he grows. This includes nitrogen-fixing plants such as clovers, vetches and trefoils. When they die, the worms eat their leaves and stems and process them into casts. The microbes get to work on this and release nutrients into the soil, thereby enabling new plant growth.

As Tolhurst related his approach to restoring soil health it struck me that his system sounded more like how woodlands work, rather than how many modern farms operate. Nutrients were being captured at every opportunity and being fed back in to enable new growth. Then he told me about his use of wood chips. 'We started composting wood chips to improve soil health by bringing in the soil fungi that live on the wood. These fungi help with phosphate release. We are mimicking woodland really. Woodland soils are the best, the most stable, which is why for centuries people have cleared them to make way for crops. But the nutrients are quickly depleted.'

He showed me a twenty-metre-long strip of wood chips in various stages of decay. Taken from local garden trimmings, the chips are used to cultivate the fungi and bacteria that will not only rot the wood and make the carbon in it available as fuel microscopic soil organisms, but will also add to the soil's ability to release and transfer nutrients to crop roots. A cloud of mist rises from above a pile of recently arrived chips. I put my hand inside. It is as if the chips had been gripped by fever, the moist heat being generated by the work of wood-feasting fungi. As we walk down the length of the heap, we reach older material that has been there for three months. The fungal threads are visible in the darkening chips. At the far end we reach the oldest part of the pile. It is cool and looks like soil. This carbon-rich material is ready to be spread among a crop of green manure. Earthworms will take it below ground where the carbon will fuel microbes and the fungi help with nutrient transfer.

While he seeks to mimic a woodland system, Tolhurst told me that his approach manipulates nature no less than farms using external inputs on a grand industrial scale. 'Farming is all about control, but it's the philosophy that's important and what we are trying to do here is to work with nature rather than against it. I am a control freak, but I am trying to control and manipulate natural systems and doing that in a way that builds soil health and stops nutrients being lost.'

The methods being used on this little farm attract scepticism, not least because of the lower per hectare* yields compared with farms using copious manufactured nutrient and pesticide inputs. I asked Tolhurst if in the face of rising demand for food we can afford to farm in this low-input manner. 'We can't afford not to,' he replies, and points out how in any event

* A square hectare covers an area of 100 metres by 100 metres. There are 100 hectares in a square kilometre.

conventional high-input farmers would not even attempt to use the poor-grade land that he is exploiting to such good effect. Their methods would be too crude for the soil here to work for them.

Then he tells me of the relative efficiency of what he is doing compared with more industrial operations. 'When you start measuring the fossil energy inputs and the importation of fertility from elsewhere, you realise just how vulnerable the whole thing is. In terms of calories of food produced compared with calories of energy needed to grow it, we are very efficient. Last year we even had a positive carbon footprint as well. We took more carbon into the soil than we released from energy use and everything else.'

Tolhurst's little eight-hectare plot enables four people to make a year-round living, with a few more part-time jobs in the summer. On a nearby arable farm growing cereals it would be about one person per 200 hectares. While that latter approach is often referred to as intensive farming, this farm is also very intensive. The difference is about what kind of intensity. In Tolhurst's case the level of intensity is not seen so much in machinery and chemicals but more in the brainpower and ingenuity devoted to nutrient conservation and improving soil health.

Nutrients from waste

Tolhurst's small-scale operation feels solid, robust and resilient. It is of a scale that works with the place it occupies. It is securely embedded in its landscape, where it complements rather than cannibalises its surroundings. The farm strikes a deep contrast to those devoted to the large-scale production of a single crop, mostly cereals in that part of Oxfordshire, and with the import of substantial quantities of manufactured fertiliser and pesticides needed to do it. These contrasting systems in the end both rely on soil, though one is improving soil condition, and the other is causing its deterioration. Even on big farms this need not be the case, however.

Dr Arnie Rainbow is the Research and Development Manager at Vital Earth Limited. His company makes money by catching nutrients from organic wastes that would otherwise be disposed of in rubbish dumps and incinerators. The company's £10m facility, opened in 2006 at Ashbourne, Derbyshire, is the largest, most advanced in-vessel composting facility in the UK. Using technology in tandem with nature the facility makes a range of organic composts, mulches, soil improvers, substrates for green roofs and other growing media.

Rainbow says that some twelve million tonnes of compostable material goes into British landfill sites each year and that this is increasing annually at a rate of about 2 to 3 per cent. As well as releasing methane, landfills take up a lot of land – in total equivalent to about half the size of the Isle of Wight – 'an appalling waste of plant nutrients, and pretty unpleasant to look at as well'. In 2009 Vital Earth produced about six million cubic metres of product.

During recent years the company has seen more and more interest in compost from farmers, which, as Rainbow says, 'is amazing given that composters once had to give it away, or pay farmers to take it. Now the ones using it point out a number of improvements. They can plough fields more quickly, by improving the structure of the soil, and they get less drought stress on the crop, less disease and of course it reduces the need for applying synthetic fertilisers and lime. It feeds the soil, encourages worms and reduces soil loss.'

Recovering organic materials in this way can also act as a hedge against price volatility in key resources. Ammonium nitrate production, for example, relies heavily on natural gas, and when gas prices go up, as they have done in recent years, so do does that of ammonium nitrate.

One farmer who uses Vital Earth's compost products is James Chamberlain. He farms 680 hectares near to Shardlow, south of Derby, where he's been producing a variety of crops and livestock for over thirty years. He's been using compost to improve the condition of his soils for some years. 'It's very important for aiding soil structure, to feed the microbes, the earthworms, and to create more pore space that holds water, and retains water more readily in drought times.'

Chamberlain has seen a progressive improvement in the quality of his soils, including their ability to hold water. He points out how compost is relatively new in farming and observes how the increased availability of such materials is down to councils setting up household recycling schemes that take food and other organic waste. 'It's only when they introduced the Landfill Tax that there was the impetus to recycle and reuse organic wastes'. Composting at small and large scale is thus a means to reduce wastes that would otherwise be buried in the ground, and a way of catching nutrients and managing costs while improving the quality of soils.

Another scalable route toward the protection of soils can be seen in research findings from the Allerton Project. This programme has been run by the Game and Wildlife Conservation Trust on a Leicestershire farm since 1992 and has trialled various methods to render the production of wheat, beans, oilseed rape and oats more sustainable, including from the point of view of the soil. I visited the project back in the late 1990s and was impressed by the work being done there to find out how to maintain and increase food output while also increasing game and other wildlife populations.

Its researchers have, for example, discovered that 80 per cent of the soil that is leaving the fields can come from the tracks used by tractors – the so called tramlines – even though they occupy only 2 to 3 per cent of the field area. Simple remedies such as ensuring that tractor tracks follow contours, rather than go up and down slopes, has been found to make a big difference. So has the use of bigger tyres that spread the weight of machinery, thereby reducing compaction.

A big reduction in the amount of run-off reaching streams has also been achieved through putting in ponds. Sited so as to intercept water running from fields, sediment and nutrients settle out and thus reduce the quantity getting into nearby streams, where it causes damage to wildlife, including the gravelly spawning habitat of wild trout.

Even with the very best of practice, however, all farms, big and small, export nutrients from the fields. No matter how successful farmers are at keeping nutrients there during growing and soil recovery, there is no choice but to let nutrients out in the crops produced. That is after all the point of farming – to turn nutrients into food.

For this fundamental and seemingly impossible-to-plug gap for nutrient loss, however, there are ways to mimic nature's economy and to replace linear nutrient pathways with more circular ones. Downriver from Tolhurst's farm there is an example of just such a thing.

Real green crap – and for profit

The Slough sewage treatment works is an unromantic place, spread beneath a major flight path for aircraft leaving Heathrow and by the side of the M4 motorway. The aroma of human waste penetrates nostrils and throat. Processing materials that most of us would rather not think about, this huge complex treats some 600,000 tonnes of raw sewage every day, in huge vats that look like oil refinery tanks. These are filled with the bubbling black discharge from millions of toilets. The tanks are a powerful reminder that all waste has to go somewhere.

Sewage has been treated on this site for about 100 years, and rather like an ancient city has accumulated layers of history that reflect successive waves of cultural and technological change. An old settling pond sits forlorn. Machinery that once aerated water is stuck fast with rust. Blue paint applied in the 1950s peels from its once modern carcass. From the most basic facilities that simply separated solids from liquids before releasing heavily polluted water back into the River Thames, the treatment works have over time adopted methods to progressively improve the quality of the water that they discharge to the environment.

The sprawling complex of tanks, pipes and ponds might not seem like the kind of place where you'd expect to find the means to achieve more sustainable farming, but in one corner, among the decaying remains of kit from decades past, there is a brand-new building that seeks to do just that. Inside the sleek corrugated steel structure is a glimpse of the future, designed to extract phosphate from sewage liquid.

I was honoured to metaphorically 'cut the ribbon' for this new installation at a ceremony that took place on a wet grey morning in November 2013. The plant is located next to an older centrifuge facility that separates solids from liquid material. A conveyer belt takes the black sludge and drops it into a vast concrete compound from where it is bulldozed into glutinous mounds. These solids are rich in nitrogen.

They will be further processed and dried before being spread on land. The liquids spun off by the centrifuge are packed with phosphate, initially extracted from distant mines and turned into fertiliser, before being used to grow food. A little bit of the phosphate here will have passed through the digestive systems of Iain Tolhurst's customers. Some may even have first got here via mummified cat bones!

Whatever its origin, the precious and depleting phosphate would make its way to the sea via rivers, where it would do damage to the aquatic and marine ecology. Not here though, not any longer at least. Instead, a lot of the phosphate is intercepted, turned back into fertiliser and sold to farmers who add it back to soil.

Thames Water's Simon Evans told me how the benefits of the new plant went beyond the recovery of a depleting resource and protection of the environment. 'Here we are producing a better fertiliser product than that coming from rock phosphate. It is purer and is not soluble and so won't run off the ground when it rains. It is only released in contact with plant roots, so none is wasted. Because it is nearly all used up by the plants it doesn't get into waterways and cause environmental problems.'

I was shown how the facility works. A cone-shaped reactor receives wastewater from the centrifuge and a chemical catalyst is added causing the phosphate to form crystals. The tiny nodules are pushed up to the top of the vessel where the water is moving more slowly, but as the crystals grow and become heavier they drop toward the bottom, where the product is harvested. Through a glass pipe at the base of the reactor the operator can see the size of the crystals of phosphate inside.

When they are about three millimetres across they are separated, sent to a drier and then expelled into one-tonne bags. As I ran my hands through a great sack of tiny phosphate pearls I pondered the profound importance of this material to the cycles of life on Earth, and how without it plant life stops.

The technology is very flexible and can help deal with some of the environmental challenges that come with more and more people living in towns and cities. The Slough plant will produce about 150 tonnes of phosphate per year, but one being built in Chicago will salvage 10,000. It is a technology that fits well with the continuing global process of urbanisation, because as people become more concentrated in cities so do the nutrients that arrive in their food.

Recapturing those nutrients as they leave again via the bottleneck of sewage treatment works makes sense at a number of levels, although as yet not everyone can see them. My visit to the new plant came a few days after the Prime Minister had launched an attack on water companies for spending too much money on environmental schemes and thus increasing customer bills. I asked Evans how he could justify two million pounds being spent on the new technology at a time when wages were static and costs for consumers rising. He told me how the nutrient-recovery facility was in fact *saving* his customers money. 'The phosphate builds up as a rock-like scale on the inside of the pipework and in a matter of months they can be completely blocked. Each year this costs us and our customers about two hundred thousand pounds to sort out on the Slough works alone.'

This fact, coupled with the revenue stream that comes with the sale of phosphate fertiliser, means that the Thames Water plant will pay for itself in under ten years, and from then on will be saving money for the rest of the time it remains in operation – around twenty-five years for most of its components. The payback time for larger plants like that in Chicago is even lower – typically three to five years. It's a genuine win, win, win, win: for nature, food security, bills and profitability. Exactly the kind of technology we need more of.

Digestion for eating

Another technology that has great potential in helping to deal with some of the environmental, economic and security challenges we face in relation to nutrients and soil fertility can be found in an industrial area on the outskirts of the fenland town of March. Behind a vegetable processing and packaging plant that receives thousands of tonnes of potatoes, leaks, carrots and onions from the wide peaty fields of East Anglia is a huge grey tank, some ten metres high and about the same in diameter. It is connected via an array of pipes to a series of shiny steel processing vessels. This is an anaerobic digestion plant and it basically mimics the kind of process that goes on in lake beds, deep ocean ooze and rubbish dumps, where organic material eaten up by microbes in the absence of oxygen results in the production of methane gas.

Produced in the large grey digester, the gas is collected in a large white dome. The gas is the same stuff that comes out of the ground as natural gas, although this bio-methane is renewable and could be an increasingly important alternative to its fossil equivalent (including that which might be produced by fracking). The gas collector dome is also hooked up to a generator, producing electricity with the heat it creates as a byproduct of that used for processing new batches of food waste before they're added to the digester.

The system is run by a company called Local Generation and in 2013 the business took about 42,000 tonnes of food waste from supermarkets, local council collection schemes and food-processing companies so as to generate renewable gas. Lee Dobinson is the company's Commercial Director. He told me that organisations generating substantial amounts of food waste were keen for his company to take it off their hands because putting it in the digester was cheaper than sending it off to landfill, due to the tax charged on disposal via that route. 'At the moment our output is just under two megawatts of electricity. In the long term, when we get up to 70,000 tonnes of waste, we'll be producing about three megawatts. The factory next door takes between a quarter and a third of our energy output, and the rest is going into the National Grid.'

When the organic material is exhausted for the biomethane it can produce, the digested residue is recovered and added to soil, in the process recovering nutrients for food production, including phosphate and nitrogen. So instead of using gas to make fertiliser, this process produces gas while making it, and it is renewable too. Dobinson highlighted to me the circularity of the technology. 'We're getting rubbish in, growing new food, the nutrients go out and then we're getting waste food coming back in, and that's running the whole process and generating a lot of power.' As the UK and world population continues to rise and as different kinds of environmental stress become more pronounced, this is another key technology we need to scale up.

Farmers have cultivated Britain's soils since the end of the Stone Age. Because of the recycling miracles worked by soils' circular economy, and nutrients derived from often far-flung sources, our land has been able to feed a growing population. We should not take it for granted that rising demand can continue to be met indefinitely, however. The way we have come to use and abuse soils and waste nutrients has left us vulnerable to external shocks, such as gas price volatility and depleted mineral deposits, and caused costly environmental impacts.

The good news is that we can do something about all this. By slowing down the loss of nutrients from farms, and by recapturing some of those that do escape, a big dent can be made in our reliance on energy-intensive and limited resources while leaving our environment in better shape. Central to doing all this effectively is soil health, including measures to stop erosion and to keep soil where it belongs – in the fields and pastures.

Soil is not the only natural system we rely on for our food. Pollinators and predators help to feed us too, as we'll see in the following chapter.

Manifesto

- **Establish an official UK Soil Institute** to monitor soil health and advise ministers across the UK on policies and decisions that affect soils.

- **Ban waste food** and other organic materials being sent to landfill (this is already law in Scotland) and increase incentives for investment in anaerobic digestion technologies so as to meet 10 per cent of gas demand from this renewable source by 2025.

- **Direct water companies** to undertake cost-benefit studies on the installation of phosphate-capture technology at their sewage treatment works.

A bumblebee at work – helping to secure Britain's food supply.

Chapter 2
Bees, bats and earwigs

£430 MILLION – VALUE OF THE SERVICES PROVIDED BY POLLINATING INSECTS FOR CROP PRODUCTION

76 PER CENT – PROPORTION OF PLANTS FAVOURED BY BUMBLEBEES FOR GATHERING NECTAR AND POLLEN THAT HAVE DECREASED IN FREQUENCY

31,000 – TONNES OF PESTICIDES USED EACH YEAR IN BRITISH FARMING

William Thatcher set up business in the village of Sandford in 1904. The cider company bearing his name remains there today, its buildings spread along the main road and between the houses that lie beneath a verdant oak- and ash-clad spur of the Mendip Hills and just next to the Somerset Levels.

At its substantial factory site where the company's products are manufactured, packaged and shipped to market, six-metre-high oak vats dating back a hundred years are still in use. Because of Thatchers' explosive recent growth they have recently been joined by a battery of 24 new shining stainless steel fermenting vessels that hold 60,000 litres apiece. This and other processing equipment enables the company to produce about 50 million litres of cider per year. When it's working at full steam the factory can fill 20,000 bottles an hour.

This iconic West Country business turned over about £60 million in 2013 and exported to several overseas markets, including the USA and Australia. Employing about 120 people directly, its supply chain of bottles, kegs, haulage and packaging supports jobs in many other firms. It is a British business success story, growing 23 per cent in 2013 – the lowest rate of growth it had seen in eight years. Behind its success is a trusted brand, a lot of technology, complex logistics, good supplier relations and business acumen. Less obvious at first sight is another even more fundamental asset: insects.

More than 500 hectares of apple trees each year supply the company with about 20,000 tonnes of fruit. For the trees to produce the apples harvested in late summer and autumn it is vital that in spring their flowers are pollinated. That mostly happens through flying insects passing from blossom to blossom, as they collect nectar and pollen to eat.

Old hedges, some of them with ancient trees, divide up the farmed landscape around Sandford. Among a mixture of pasture and crop fields are orchards with tens of thousands of apple trees, some young and slender, others old and gnarled. Bursting into leaf and flower under a dramatic showery May sky, I'm shown hundreds of different apple varieties, with pink fragrant eruptions peaking on some trees, fading on others and on a few still emerging. It is a critical time.

Ecological glue

The insects that pollinate the apples, which make the cider that gets sold to sustain a multi-million-pound business and jobs, are a kind of ecological glue. Take them away and the system can start to fall apart. For example flowering plants either stop producing seed and fruit, make much less of it or change the quality of what they do produce.

When it comes to conserving the natural environment, all this is very important to understand. It is also crucial to food (and drink) security. About two thirds of the world's crop plant species rely on animal pollination, and to that extent bees, butterflies, beetles and all the other species that help plants to complete their life cycles ensure that we continue to eat.

The loss of pollinators has in some places already emerged as a major challenge for farming, for example in parts of California (where bees are shuttled around the state) and in some regions of China (where pollination is done by hand, using feather dusters). While that point has not yet been reached in Europe, the rapid and ongoing decline of pollinators is nonetheless quite rightly regarded as a strategic threat to food production. This is not surprising when you consider just how quickly many of those creatures have declined.

While most species of flowering plant rely on animals (mostly insects) to move their pollen between flowers, some use the wind, including the grain-producing wheat, barley and maize that occupy so much of our farmland. Even so, animal-pollinated crops, such as oilseed rape, apples, pears and strawberries, comprise about a fifth of the UK's total cropped area. Many pasture plants, such as clover, also rely on bees to reproduce, so there's some dependence for livestock too.

Since 1980 there has been a reduction in wild bee diversity in most British landscapes. In line with the progressive homogenisation of our environment, those species that are specialists and dependent on particular kinds of habitat and/ or food plants have fared worst, with those able to tolerate more general conditions doing relatively better. Many other insect pollinators are also known to be in decline, among them hoverflies and butterflies.

Behind all this has been a damaging loss of habitats rich in wildflowers. Wildflowers are primarily found in semi-natural grasslands, mountains, moors and heathlands, woodlands and coastal margins, and some urban habitats, such as parks, gardens and roadside verges. In all cases, and for the variety of reasons described in this book, there has been a decline in the quality of these ecosystems.

This is a problem for farming because the pool of pollinators supported by the plants found in these places spills out into agricultural areas. Flower-rich wildlife habitats are not, therefore, only very nice to look at and of huge ecological interest and value for their own sake, they are also an essential part of what might be seen as another aspect of the 'green infrastructure', in this case underpinning national food security.

And while pondering the dependence of many insects on wildflowers, it is important to remember that the wildflowers depend on the insects. A high proportion of wildflower species studied are pollination-limited. This means that when there are no pollinators there are no seeds, over time fewer flowers, and then fewer pollinators. In line with the reduction in bee diversity, animal-pollinated plants have declined in the UK more than self- or wind-pollinated species. For example, some 76 per cent of plant species favoured by bumblebees for gathering the nectar and pollen they need for food have decreased in frequency.

Decline in particular kinds of plant that have lost their pollinators can also lead to the decline of species dependent on their leaves, seeds and fruits for food. Changes in plant communities can then cause knock-on impacts on soil health, water quality and pest regulation, with disruption to pollinators leading to a spiral of decline in an entire system.

Professor Jeff Ollerton is based at the University of North-ampton where he is an expert in pollination and its role in ecosystems. 'About seventy-five per cent of our native plants require insects as pollinators,' he says. 'If we didn't have the insects to pollinate them, then our populations of native wild-flowers would eventually die out.' It's not just little plants that have this dependency: about 60 per cent of our trees are insect-pollinated.

The loss of semi-natural habitats (those with mostly native species but in a human-modified state) across the UK since the mid-twentieth century has been a major factor behind pollinator loss. The ploughing of meadows, clearance of woodlands, drainage of wetlands and removal of hedgerows are among the changes that have caused it. Most of these losses are down to changed farming practices, backed over decades with tens of billions of pounds of tax-funded subsidies.

Ollerton says that a substantial number of pollinators are already lost. 'Twenty-three species of bees and flower-visiting wasps have gone extinct in Britain since about 1850. This maps very closely against the large-scale changes in agriculture. There's a very strong correlation there.' He adds how the distribution of many species that have survived are in serious decline. 'We've got about two hundred and fifty species of bee that are native. A significant pro-portion of those have declined in relation to distribution. We've got some bumblebees, for example, that were once widespread across the country and which are now being pushed up into the Scottish Highlands and Islands and out towards Wales.'

Ollerton points out how this decline in diversity, although not yet causing visible impacts, leaves farming more vulnerable. 'If all of our pollination is reliant upon just a small number of species and something happens to those species – for example there's a lot of concern about diseases in bumblebees populations – and they die out, then there's nothing to fill that gap. That's one of the values of diverse insect populations, a sort of insurance policy.'

The changes in farming practices have been huge and profound. 'If you go back into the nineteenth century, set aside was very common, where farmers would leave a field to go fallow, or maybe sow a crop of something like red clover and then dig it in to build up the fertility of the soil. By the time we get into the twentieth century, when inorganic nitrogen fertilisers started to be produced, suddenly there's no need to have those fallow periods or to use clovers and you can intensify production and continue using fields all year around. Therefore that fallow field, that habitat, that resource to be used by pollinators has disappeared. That was a massive landscape level change.'

Chemical weapons

In the wake of the virtual disappearance of semi-natural grasslands, the extinction of wild honeybees, decline in bumblebees, beetles, butterflies and others, some assume that domesticated colonies of honeybees will fill the gap. Here too though there is cause for concern.

For a start there are far fewer hives kept than in the past. The UK's high point for bees was in 1949, when in England alone there were 87,000 beekeepers and 465,000 colonies. In that post-war period the popularity of beekeeping was less to do with pollination and more about sugar rationing. When that particular restriction was lifted in 1953 beekeeping for honey became less necessary and by 1970 the number of beekeepers in England and Wales had dropped to about 32,000, running about 158,000 colonies between them.

Beekeeping has become more popular again since then. It is estimated that during 2009 there were 40,000 beekeepers and 200,000 colonies, including approximately 300 commercial beekeepers between them managing 40,000 colonies. The recent surge in beekeeping has, however, been accompanied by a rise in bee colony collapse, a process leading to the death of a hive. There are a number of reasons why this has happened. Often it is linked to the parasitic *Varroa* mite that infests hives and increases the bees' susceptibility to harmful diseases.

On top of these infestations, however, are the effects of pesticides. Around 31,000 tonnes of chemicals are used in farming in the UK each year, to kill weeds, insects and other pests that attack or damage crops. There is surprisingly little control over how these chemicals are used (though they are, of course, banned in organic farming) or in what quantities and combinations. The chemicals that are so liberally doused across the land affect both wild and domesticated pollinators and include consequences that range from outright death to changed behaviour and increased susceptibility to disease.

In recent years there has been intense controversy over the role played in both wild and domesticated bee decline by a relatively new class of pesticides called neonicotinoids. Designed to affect the central nervous system of insects, these substances cause behaviour change, paralysis and death – all of which must obviously damage the work of pollinators. Studies showing how these chemicals can accumulate in the pollen and nectar of treated plants reveal how such creatures are at particular risk.

In experiments that exposed bumblebees to a widely used neonicotinoid substance called imidacloprid at a level the insects would experience in visiting a field of treated oilseed rape plants, the insects were found to forage less effectively, colony size shrank by about 10 per cent, and most worrying of all, they lost almost all their ability to produce queens. And without queens new colonies can't be established.

Evidence like this led the European Commission in 2013 to propose temporary bans on three neonicotinoid products. While most European member states backed the move to suspend use of the chemicals in question, the UK waged a ferocious campaign to keep them in circulation. Backed by pesticide companies, including Syngenta and Bayer Cropscience, the UK government did its best to resist the proposed suspension.

The companies' argument (ventriloquised by British Ministers) was that the 'smoking gun' research linking bee decline and neonicotinoids was not conclusive. Part of the challenge for the scientists investigating neonicotinoids arises from how effects can occur at quantities so tiny that they are virtually unmeasurable. For substances that cause profound effects on insects in concentrations measured in parts per billion, coming up with definitive and direct proof as to their impact is, to put it mildly, difficult. On the other hand, marshalling actual proof there is no effect is equally challenging.

That said, the evidence as to the negative effects of such chemicals grows only stronger and now includes research that indicates impacts on other wildlife. For example findings published in July 2014 established a relationship between the use of imidacloprid and declining bird populations.

Scientists looking at an intensively farmed part of the Netherlands found that population trends among insectivorous birds were more negative in areas where higher concentrations of that pesticide were detected in the environment. This relationship appeared only after the introduction of that chemical in the mid-1990s and was still apparent after correcting for the kinds of land-use changes that are known to affect bird populations. The authors of the research said that laws governing pesticide use should take account of such findings.

The law is of course a matter of politics as much as of data and evidence, with the decision as to what to do in the light of findings like this a question of balancing risks and benefits and assessing pros and cons to hopefully reflect the greater social good. Sales of neonicotinoid pesticides earn many millions of pounds for the companies that make them and in so doing generate tax receipts for governments, and it seems that for reasons of economy the British government policy saw taking risks with ecology as more desirable than a temporary restriction on one part of one business sector. In siding with the chemical companies the UK government's stance basically amounted to disregard for the so-called 'Precautionary Principle' (the idea that when doubt exists the benefit should be with protecting the environment) in favour of decisions to boost GDP in the short term, whatever the longer-term costs.

Despite the best efforts of the companies and the UK government that backed them, the European Commission's proposals did in the end prevail, and some neonicotinoid chemicals were at least temporarily taken out of use. Researchers will try to assess what happens to bee populations in the wake of the ban, although with a suspension of just two years it may be difficult to reach a clear conclusion.

While it is too early to pass verdict on the effect on bees, some farmers have reported pest infestations in oilseed rape fields that they say can't be controlled because of the ban, leading them to make loud calls for it to be lifted. While farmers' short-term difficulties must be taken seriously, I only hope that commentators and ministers will take the discussion beyond the simple binary of pesticides versus no pesticides, and instead look at alternative approaches in the form of, for example, different crop rotations and more natural pest management (a subject we'll come on to).

Costing the essential services

Whatever the combination of reasons for wild pollinator decline and the collapse of honeybee colonies, there's no doubt that domesticated honeybees play a far smaller role in providing pollination services than wild species. Based on an analysis of the area of crops needing insect pollination and the recommended densities of hives needed to pollinate them, the number of registered hives in the UK has been calculated as sufficient to supply only a fraction of the pollination services we need.

Dr Tom Breeze, a Research Fellow at the University of Reading, has spent a lot of time looking into the contributions made to UK farming by pollinators. He says that it used to be assumed that whilst other species could pollinate crops, honeybees were the only ones that mattered. 'Now we know that even under the most generous assumptions they provide only a third of the pollination services that we require. Other species are definitely filling the gap.'

That's a big gap, and no matter how many honeybee hives are established to fill it, for certain crops, including strawberries, tomatoes and peppers, it won't help because bumblebees are their main pollinators. For some others, including field beans, apples and raspberries, honeybees are not as effective pollinators as wild insects. There are, therefore, plenty of good practical reasons why we should be concerned about the ongoing decline of natural pollinators.

When recent estimates as to the economic contribution made by pollinators are taken into account the case for action to reverse the fortunes of the wild ones becomes even more compelling. The figure that Tom Breeze came up with, based on UK crop production in 2007, suggests that pollinating insects provided services that underpinned crops with a market value of about £430 million.

The total contribution to the economy is even bigger than that estimate, however. Breeze's calculations don't include the value of pollinators in sustaining forage crops, such as the clovers that support livestock and thus much of the meat and dairy sector. They don't include small-scale food growing, such as that in allotments and gardens. Nor ornamental flower production or the UK's multi-million-pound seed-producing business. Nor the value added to the wider supply chains of companies like Thatchers. And then there is the contribution that wildflowers and insect-pollinated trees make to our quality of life and wellbeing (which is a real economic factor, too, as we shall see in Chapter 8).

While it is unlikely that we'll imminently see a total collapse in the services provided by pollinators in the UK, several studies have demonstrated that when they become locally depleted reduced crop yield and quality can be one result. Should they all be eliminated, Tom Breeze says that his team estimated that it would cost 'around £1.8 billion to replace insect pollinators in the UK'.

In addition to financial figures relating to the broader value of pollinators, there are published studies that have set out to quantify wild pollinators' contribution in relation to particular crops. An interesting piece of research conducted at Simon Fraser University in British Columbia studied farms in Alberta producing oilseed rape. They found that farmers who planted their entire farm would earn about $27,000 in profit whereas those who left a third unplanted for bees to nest and forage in would earn $65,000 on a similar-sized area. Talk about counter-intuitive.

One way to avoid taking the trouble to keep pollinator populations healthy, and the costs of restoring them once they're gone, would be to take British farming in a different direction. For example we could grow only wind-pollinated crops. While this would have the advantage of making our farming sector less dependent on insects, it would require a change in food choices. UK consumption would either have to shift away from fruit and vegetables and more toward cereals, or a greater reliance would have to be placed on imported pollinator-dependent foods.

At a time when official health advice recommends a higher proportion of fruit and vegetables in our diets (recently raised from five portions a day to seven), keeping pollinator populations healthy will need to be a part of the same policy. And when food-price fluctuation is expected to become more volatile it would seem an intelligent choice to maintain our capacity to grow as much of our own as we can.

I am not alone in concluding this. In the summer of 2014 the House of Commons Environment, Food and Rural Affairs Committee said that more should be done to increase the proportion of homegrown fruit and vegetables that we consume. The committee pointed out how in 2012 we imported some £8 billion-worth of fruit and vegetables, producing only 12 per cent of our own fruit and just over half our vegetables.

In this context the Committee said that: 'While the UK may be food secure at present, it would be unwise to allow a situation to arise in which we were almost entirely dependent on food imports given future challenges to food production arising from climate change and changing global demands.' Changing the proportion of what we import is, of course, not just about trade relations, it is among other things linked with the state of Britain's pollinators, begging the question of how best to maintain and restore them.

Rebuilding the nectar networks

There are some obvious priorities. The first is to render landscapes more pollinator-friendly by, for example, ensuring more field margins are rich in wildflowers, protecting and recreating flower-rich meadows and conserving and restoring woodlands, wetlands and other insect-supporting ecosystems.

That practical and cost-effective steps can be taken in this direction is underlined by many examples across the country.

One farmer who's deliberately restoring pollinators is the compost enthusiast James Chamberlain, whom we met in the last chapter. 'I've sown mixtures of cornflower, corn marigold, corn camomile, crimson clover and red clover,' he told me, 'and we get four months of flowers, instead of just one month from your oilseed rape. This year as the oilseed rape was going off the wildflower strip kicked in, and that has been buzzing. It's my second year doing that and I have got strips like that dotted around the farm because I can see the benefit of it.'

This kind of simple change is not only being adopted by individual farmers but also by major brands supplied by many growers. One is Ribena – a company whose origins lie in an official government investigation back in the 1930s to identify homegrown sources of vitamin C. This proved to be a highly strategic question and was soon linked with a key aspect of national food security. Blackcurrants were found to be really the only good source that could be grown in our climate and made into a stable concentrate. Thus in 1938, a year before Britain was plunged into war and oranges, lemons and other sources of this vital vitamin became very hard to come by, the manufacture of blackcurrant syrup began. Taking its name from the scientific name of blackcurrant, *Ribes nigrum*, the entity set up to do this has been in the same business ever since.

Without insects the supply of the 11,000 tonnes of blackcurrants the company needs each year to make its product would not be secure. Rob Saunders, who speaks for the company, said that Ribena has come to appreciate this dependency and is taking steps to conserve the insects that enable Ribena to stay in business. 'We made contact with the Wildlife Trusts and others and put together a six-point plan that was applicable to all growers, and that was thought to be meaningful on most sites, things that were relatively easy to achieve and to implement, but that would have the maximum habitat and biodiversity impact.'

This included recommending how field margins were managed so as to keep rough grassy areas close to the blackcurrant bushes, thereby helping to maximise pollination by bumblebees. Hedges were improved for their wildlife value and more trees were planted. Forty or so growers supply Ribena with blackcurrants from thousands of hectares of bushes, so cascading these kinds of improvements across its network will make a difference to a substantial area of farmland.

Another business that has understood the commercial relationship between pollinators and profit is Thatchers Cider. I visited the orchards that supply the company with Martin Thatcher, William Thatcher's great grandson. We walked among some of the thousands of apple trees, including an orchard with 458 varieties of apple salvaged from the government's Long Ashton Research Station when its funding was axed. 'A living library,' he calls it.

On top of efforts to keep the huge genetic asset of old apple varieties going, Thatcher told me about some of the steps the company has taken to keep the vital pollinator populations healthy. 'We've done a lot planting of wildflower seeds because the thing about bumblebees is that whilst it's great for them when the blossom is out, they need a sustained food source throughout the year.'

Working with the Bumblebee Conservation Trust, surveys have been undertaken to assess the populations of these vital insects. They've also provided advice on the mixes of wildflowers that will provide pollinators with a continuous source of food during months they are active, so that they'll be there again next spring.

Thatcher tells me about the importance of hedges for bumblebees. 'They like the north side of hedges because on a hot day in spring if they're facing south it can bring them out of hibernation too early. They need to stay cooler for longer.'

The philosophy that underpins this pollinator-friendly approach is entirely practical. 'We're keen to work nature into the equation and give it every chance to succeed. If you use too many chemicals it just doesn't make sense, either environmentally or financially. You're just creating a problem for the future. What we're trying to do is to keep nature on our side. Nature needs a bit of a hand at times, especially in this situation. We're aware that we can't rely on honeybees.'

Among Thatchers' future plans is an ambition to entice more species into the orchards. 'There are a few very rare bumblebees not far away and we're keen to encourage them up here. We can't pick them up and move them but we want to create the right environment.'

On top of more careful chemical use and the enhancement of pollinator habitat in farmed landscapes there is also an important role for protecting those last areas of wildflower-rich habitats that remain, and not only the lowlands but also in those apparently wild-looking areas in the North and West of Britain.

The windy uplands of Northern Ireland's Belfast Hills present a case in point. It looks green and lush up there but areas supporting the once wide range of natural pollinators are now very few. Slievenacloy is the name of a nature reserve that translates in English to 'Hill of Stones'. That gives a clue as to why the flower-rich meadows survived here when nearly all the others nearby were 'improved'. Although it was far from ideal for agriculture, even this place came under threat, until it was officially protected and brought under sympathetic management by conservation charity Ulster Wildlife.

In summer tens of thousands of butterfly orchids bloom here. These rare plants are part of an ecosystem that not only supports rich flora but the many kinds of insects that depend upon it, including very rare species such as the grey mountain carpet moth. Through a combination of sensitive grazing and control of bushy scrub, Slievenacloy's diverse wildlife is being conserved, in the process providing a reservoir of natural pollinators. The rich mix of plants sustains a rich variety of insects, which sustain the plants, which sustain the insects, some of which support farming and food production.

There are many such areas dotted around the UK, although what we have now is only about two per cent of what we had in the 1940s. Keeping what's left in good shape can be increasingly seen not only as vital conservation work but also as one plank of food security. This place is now under the kind of management that will perpetuate its wildlife, but at the same time raises big questions as to how what remains might be kept and how to restore some of what's been lost.

So it is that the kinds of simple ecological enhancements made by businesses like Thatchers and Ribena, tighter official restrictions on pesticides, and the protection and restoration of flower-rich habitat, can make a difference. If this is done well and at greater scale we might find other benefits for food production coming in its wake, including more sustainable pest control.

Pest heaven

While for many decades now we have become used to waging chemical warfare on some of the animals that eat and spoil our food crops, there is increasingly good reason to see benefits in a more sophisticated approach – one that finds allies among many of the creatures that we have hitherto been killing.

For fast-breeding small insects that feed on crop plants, such as aphids, the planting of large areas of a single species can be bug heaven. Feeding, laying eggs, eggs hatching, babies growing, laying eggs themselves, eggs hatching, their world, when comprised of a large field with a single crop, is in a sense unlimited. Numbers can mushroom. With multiple generations during a single season, exponential population growth not only causes direct damage to crops but also spreads plant diseases.

Our chemical weapons generally keep pace, but at some considerable expense and with serious collateral damage, seen in the loss of wildlife diversity beyond the pest species targeted for termination. This is not only seen in the loss of pollinators. Another cost appears in household and business bills, as expensive technology needs to be installed to remove chemicals sprayed on fields from our drinking water supply. Pesticides sprayed on British fields have travelled right around the world, to accumulate in among other places the blubber of polar bears, whales and penguins, and indeed in our own body fat.

Further downsides arise from the biological counter-measures deployed by pests in response to our molecular arsenal. Our chemicals might be increasingly sophisticated but they're not as sophisticated as nature. Aphids giving birth at the start of the growing season will be succeeded by five more generations at the end of it. With evolutionary forces at work on each batch of newborns, natural selection favours individuals that have a genetically determined tendency to withstand the poisons we unleash. The shorter the intergenerational period the quicker the natural selection works, and that is why over recent decades literally hundreds of pest species have evolved resistance to many of our chemicals.

One way forward is to use more chemicals and to develop new ones. Another is to alter farming practices and to take steps to enhance the number and diversity of the pests' natural enemies. Research has demonstrated, for example, how increasing beetle and spider populations can be especially beneficial in protecting yields. This is why fields that have a high proportion of rough grassy margins and hedges compared with the area of crops, and those planted with perennial crops, are associated with lower pest establishment.

One person who knows this very well is Iain Tolhurst, whom we met at his farm by the River Thames in the last chapter. Not only does he mimic and nurture nature in how he builds soil fertility and maintains pollinators, he adopts a similar approach for pest control. You will remember how his farm has been taken in a different direction of travel to the kind of input-dependent monoculture that official policies, subsidies and agrochemical companies encourage. But as he walked me around his fields he told me how he has 'very little problem with pests and diseases'.

He pointed to a beetle bank, a strip of land about a metre wide planted with various non-crop plants. Predatory insects and spiders live in there, he told me. The proportion of banks to field is carefully worked out to ensure hunting insects can travel into the centre of fields to eat pests. Next to the beetle bank the green manure, comprised of vetch, trefoil and grasses, among other things, is also an important insect habitat. Not only does it help to rebuild the fertility of the soil, it brings in more insects, and helps to reduce soil erosion.

In one plot maize stems were left standing. The cobs had been harvested and sold weeks ago, but the stems and leaves remained. This was to attract insects, Iain explained. 'If we took this away we'd lose a range of predatory insects that will eat pests next year. Earwigs are one kind. They eat aphids and red spider mites, so are important. There are also parasitic wasps. We get five or six species here, and they are important for the control of different moths. Leaving stems and other crop residues standing creates a habitat for these beneficial creatures to complete their life cycle. We are actively encouraging diversity. With that we don't need chemical control.'

The hedges that surround the fields, some of which Tolhurst planted himself, connect with the beetle banks, thereby further enriching the farm habitat for species that will help control pests. It took him a while to work out the benefits of all this. 'When I started I was like everyone else. I wanted to get rid of all the insects. I soon realised that not only is that not achievable, it's not desirable either.'

The control of pests is helped by the huge diversity of crops he grows. Unlike in monocultures there is no single massive opportunity for rapid expansion by any one pest that relies on one kind of plant, and with careful rotations he is doubly able to pre-empt pest population explosions.

Tolhurst is convinced his methods are sound, and the fact that he produces a lot of food with no pesticides says something important as to the credibility of his views. His experience underlines the importance of understanding how to better manage ecosystems to control pests through detailed long-term field studies. One person who does that is Dr Michelle Fountain, Research Leader in Entomology at East Malling Research in Kent. She investigates the ways in which natural predators can help to control crop pests.

She agrees with Tolhurst as to the value of earwigs, at least in relation to the benefits they provide in protecting tree fruit crops. 'They eat pretty much anything – fungi, leaves, pollen, but they'll also eat lots of early aphids and pear suckers, which is normally a massive problem because it's now resistant to most of the insecticides.' By hitting the aphids early in the year the earwigs help stem the potentially massive population growth that would otherwise occur as the growing season progresses.

Fountain explained how considerable benefits can be derived from encouraging a group of insects called anthocorids. 'We've done some work to show that if you enhance the hedgerows around pear orchards, you can increase your anthocorid numbers.' Once they have eaten all they can find in the hedges they move into the orchard trees where they prey on the pear suckers. 'Willows, hawthorns and nettles are really good for increasing the habitat for useful predators.'

As well as having researched the benefits of pest predators, Fountain has also come to understand some of the dangers that can accompany chemical misuse. 'For starters, it's costly and secondly if you put something on that's not necessary you might actually be doing more harm to the crop than good, in terms of knocking out beneficial predators and parasites.'

Many growers are aware of this but much more could be done in supporting them with better advice. Considering how much of the information farmers receive is from the major agrochemical companies, input from more independent experts would be positive.

This might include basic ecological guidance as to how best to ensure the habitats needed by pest predators (and pollinators) are present in farmed landscapes. Hedges, beetle banks, areas of rough grassland and patches rich in wild plants are the kinds of features that clearly help. It might also include the results of emerging research based on more detailed understanding of the life cycle of pest species.

Take ants and aphids. Some species of ants protect aphids because they 'milk' them for a sweet substance called honeydew. Michelle Fountain told me that 'Within apple orchards there's a particular problem with ants tending aphids in the trees. They try to defend the aphids because they're getting the honeydew, and they defend them against things like hoverfly larvae and lacewing larvae and ladybirds.' Knowing this basic ecological fact has led researchers to look at how the aphids might be made more vulnerable by providing the ants with an alternative food source to the aphids' honeydew. Comparable to distracting guard dogs away from sheep while wolves attack, it could prove an effective approach, especially at a time when aphids have growing resistance to some of the chemicals we deploy against them.

Integrated approaches that work *with* nature can be effective at a larger scale and on non-organic farms, too. In 2000 the RSPB acquired Hope Farm at Knapwell near Cambridge. The idea was to see if was possible to farm profitably and at the same time to reverse the decline of farmland wildlife, including the pollinators and natural pest controllers that can play beneficial roles in food production.

By adopting a series of simple measures the habitats needed to sustain different wildlife populations have been enhanced. As I walked around the farm on a gorgeous spring day with Martin Harper, RSPB's Conservation Director, he explained to me the approach that had been taken there in trying to restore wildlife populations while turning a profit from farming.

Across one field is a long strip of rough grasses – a beetle bank, like those created by Iain Tolhurst. This was one of the first environmental measures RSPB installed on the farm. Around the edges of the fields a rich diversity of wildflowers are growing. 'We have a special pollinator mix in the margins here that are brilliant for bees, butterflies and other insects,' Harper says. While we talk, Orange-tip and Brimstone butterflies flutter past. The former lay their eggs on the garlic mustard that has just come into flower while the latter seek out buckthorn bushes in the hedge. Not only are these insects very beautiful, both are pollinators. Harper tells me how in the early days RSPB scientists undertaking wildlife surveys found 'atrocious' results. 'Our surveyors would maybe see one Yellowhammer per year, but now we're averaging three-figure counts per year and more birds here in winter than we do in summer.'

As well as encouraging birds and insects, RSPB is looking at the wider environmental impacts of farming. 'We've got trials for managing water pollution. You can see the little reed-beds that have got flow-meters in there that are looking at what's actually coming off the farm. To stop run-off we're trying to keep the tractor tracks running across the slope rather than down.'

I ask him why more farmers aren't embracing the kinds of simple steps taken at Hope Farm. 'It may be an aesthetic thing,' says Harper. 'Farmers like their nice smooth wheat fields.' In changing farmer behavior he believes that there is a need for incentives and regulations, but points out how 'We've gone into reverse at a time when farmland wildlife is bouncing along the bottom, pollinators are in decline, and only twenty-five per cent of our water catchments have good ecological status. There's no way under the current plan that we're going to get the big uplift that we need.'

Harper points out the huge potential that comes with new technologies, including Geographical Positioning Systems fitted to tractors, that enable a very precise optimisation of farmland, including the much more efficient use of pesticides and fertilisers. 'Under new farming systems, you can basically program your farm, where you can maximise your production and where your less fertile bits of land can be put into wildlife habitat. The technology is there. You just need a bit of advice.'

Alongside shifting financial incentives and the potential benefits of benign technology, the idea that advice to farmers might be key to the restoration of natural values is a message that I heard time and again. When I visited that meadow in the Belfast Hills I did so with Jennifer Fulton, the Chief Executive of Ulster Wildlife, the charity that looks after the place.

I asked her what she thought was the best way forward in maintaining wildlife-rich farmland. After all, I said, protecting little areas like their beautiful reserve at Slievenacloy was probably not, on its own, going to work in maintaining healthy pollinator and pest-predator populations across the countryside.

She explained how she'd found that many farmers were sympathetic to the kind of conservation goals her organisation is promoting but because of a lack of knowledge they simply don't know what to do. In addition to protecting the last remnants of plant-rich systems, including through properly targeted financial incentives for conservation-friendly farming, she too believes that progress could be made through better-quality advice for farmers.

If at this point anyone is in any doubt as to the reality of the opportunity at hand, take note that during the decade and a half that RSPB has managed Hope Farm its profitability has increased while the bird population has *tripled*. The birds are, of course, but one indicator of the health of the whole system, including that of the pollinators and other beneficial insects. The message is clear – it is not always wildlife *or* food, it really can be both.

It would be logical if lessons proved by among others RSPB and East Malling Research were more effectively reflected in official farm policy, not least in helping to keep open our options as the climate changes. It is for example believed that a 2 degree centigrade average temperature increase could lead to an extra five generations of aphids each year, not only swelling their population but also speeding up the rate at which they evolve resistance to new chemicals.

The earlier appearance of aphids in spring can be expected to increase the damage they cause. With warmer temperatures, planting dates for potatoes and sugar beet are on average getting earlier, but are not advancing as fast as the first appearance of aphids. Aphids may thus start to arrive when crops are at an earlier and more susceptible growth stage, leading to bigger impacts on yield later. With all this in mind it strikes me that learning how to make more of beetles, spiders and wasps would be a very good idea in helping to render farming more resilient in the face of the profound changes that likely lie ahead.

It's not only predatory insects and spiders that can help, either. So can bats and birds. There is very good research to show how Great Tits nesting in an orchard can improve yields by eating thousands of fruit-harming caterpillars and Martin Thatcher has a hunch that as well as a considerable contribution from wild bees, his business is getting a lot of value from bats. They roost in the big old oaks in the hedgerows and hunt among the apple trees at dusk and dawn when moths are flying. While he doesn't yet have any scientific research to back up his theory, work from the United States suggests that the pest-control benefits provided by bats in that country are considerable.

Findings from a research team led by Gary McCracken from the Department of Ecology and Evolutionary Biology at the University of Tennessee published in 2011 found the economic costs arising from the decline of bats in North America to be roughly $22.9 billion dollars per year. That figure was calculated on the basis of the crop-protection role they perform through eating pest insects and the decreased need for pesticides. I don't know of any comparable studies looking into the pest-control functions played by British bats, but it would seem logical to assume that having more of these furry flyers in our environment would be a good thing.

Considering all this it seems the value contributed by wildlife to our food remains, as far as most policy and practice is concerned, generally underappreciated. On the positive side, however, it does mean we have a big opportunity to gain more value and to improve national food security through protecting and restoring the systems that sustain natural diversity.

Exactly the same can be said about the situation in the sea.

Manifesto

- **In order to boost pollinator** and natural pest-predator populations adopt official targets for the restoration of flower-rich habitats in farmed landscapes. Begin with the planting of 5,000 kilometres of new high-quality hedgerows and 5,000 hectares of new meadows, while improving the quality of existing habitats. Aim to achieve this by 2025 through more resources for farmer advice and conditions linked with how farm subsidies are paid.

- **Adopt a precautionary approach** in relation to pesticides that are suspected of causing serious impacts to populations of wild pollinators and natural pest predators.

- **Increase research funding** into Integrated Pest Management, so as to better understand how wildlife can be integrated with crop protection. Use the research findings to draw up guidance for farmers on strategies to protect crops with fewer chemicals.

Bluefin tuna were once found in the North Sea. This one was caught at Scarborough in 1935.

Chapter 3
Have our fish had their chips?

THREE AND A HALF TIMES – THE AREA OF SEA AROUND BRITAIN'S COASTS UNDER OUR TERRITORIAL CONTROL COMPARED WITH THE LAND AREA

HALF A MILLION TONNES – SEAFOOD TAKEN EACH YEAR FROM THE OCEANS AROUND THE UK

£2.73 BILLION – SIZE OF THE BRITISH SEAFOOD MARKET

Like most Brits I love fish and chips – especially cod. In pursuit of a fresh cod dinner, I have spent many a long evening standing on beaches and piers, rod and reel in hand with line cast into water where these wonderful creatures might dwell. Once I actually caught one. Having walked out into the sea in a pair of chest waders, I cast my line as far as I could from shore, so that my hook, baited with a tempting strip of mackerel, reached the deeper water where I was told cod were sometimes caught. It was nearly dark and I could barely see my float in the deepening gloom. Then it bobbed and slid under. I tightened the line and straight away felt the strong resistance, bumps and streaking runs of a sizeable fish trying to escape.

It took a while to get ashore, but when I finally had the fish on the beach I could see it was indeed a nice cod. Not a huge one, but at about three kilos big enough to take back for dinner. I was on the island of Møn on the Baltic Sea side of Denmark, where the cod populations were, like those in the North Sea, in a state of crisis, and where my achievement felt no less than had I caught one from a British beach. The pleasure was all the greater later that evening when deep-fried within a couple of hours of being caught, that cod made a meal to remember.

My hunter-gatherer moment with the cod was a reminder that seafood is really the only wildlife that most of us still regularly eat. Nearly everything else is cultivated in soil or reared on pasture (or in sheds). Even fish is increasingly supplied from farms, with more now produced worldwide in captivity than is caught from the wild. But wild fish is still a vast resource, making a valuable contribution to healthy nutrition and sustaining jobs across the world, including here in the UK.

The seas under the UK's territorial control extend across some 867,400 square kilometres, which is about three and a half times our country's land area. This vast expanse of saltwater hosts the widest range of marine habitats of any European country, supporting a diversity of marine life, much of which is commercially and nutritionally very valuable. That value has underpinned the economies of coastal communities around the British coast for millennia.

Over the last century it is species living near to the seabed, including cod and haddock, which have supplied the biggest catches, although their stock sizes have become much depleted since peak landings made during the 1960s. In the case of North Sea cod severe fishing restrictions had to be imposed to prevent what was feared could be a total collapse.

This struck a nerve for the British, considering how cod and chips is such a part of our cultural identity. Quite a few people stopped buying it, believing quite rightly that stocks were in dire trouble. What came across less strongly than overfishing, however, was the state of the cod's habitat, for as is the case with bees, this economically valuable species does not exist in isolation, but is the product of an entire complex system.

Some cod travel hundreds of miles, including between the southern North Sea where they spawn and the northern North Sea where for much of the year they feed. Studies into cod movements have revealed how some go much further. In June 1957 a number of fish were tagged in the central North Sea. Two were recaptured in the Faroe Islands that December, while in 1961 another one of them was caught off Newfoundland in Canada. No matter where they are though, cod need food. They are carnivores and have a varied diet, with older cod eating predominantly fish and younger ones up to two to three years old relying heavily on other animals, such as worms, shrimps and crabs that they hunt on or near the seabed. Cod of this age comprise a large proportion of the overall catch, so their food supply matters to the human economy. In other words, those of us who like to eat cod (or indeed any wild seafood) are part of an ecological web that connects us to the health of the oceans.

Trawling for solutions

Cod (and most other fish we eat) are caught in trawl nets that are dragged across the seabed. These often cause damage to the habitat of the crustaceans, worms and molluscs that are among the main food sources of target fish. Damage to features such as reefs and seagrass beds can be especially troubling, as this not only disrupts food webs but also removes the spawning habitat of some commercially valuable species.

While from the shore the sparkling sea can seem like an unspoiled wilderness, below the waves great changes have occurred, especially because of trawling. A piscatorial atlas of the North Sea published in 1883 included a map of the distribution of seabed types. It depicted a huge 20,000-square-kilometre oyster bed and showed a number of extensive stony reefs beneath the sea along the south and east coasts.

Following more than a hundred years of intensive trawling and dredging, most of these prime fish habitats have gone. The oyster bed was last harvested in 1936 and by 1970 was completely obliterated. In addition to being important for fish it also provided a purification function, as the huge area of filter-feeding molluscs cleaned the water as they sifted suspended particles in the water column.

In addition to the loss of once widespread marine habitats there have been profound changes in the mix of species living in our seas, with large predatory fish especially reduced. Big mature cod are far less frequent than a few decades ago. Other species, such as the population of Bluefin Tuna that once lived in the North Sea, are now extinct. Fish in excess of 300 kilos were regularly caught around the coasts of Lincolnshire and Yorkshire, the voracious predators gaining such enormous dimensions by feeding on the huge shoals of herring once found there. The disappearance and reduction in big predators like these has had profound knock-on effects, equivalent to the eradication of bears and wolves on land. Entire food webs have been affected.

Major damage is also being inflicted on marine environments by pollution coming from the land. Nutrients and sediments from agriculture, and to an extent sewage works, are big issues, causing not only mortality among fish and shellfish but also the loss of their food species, some of which are very sensitive to pollution. This problem can in part be addressed through more advanced water treatment, for example with the kind of technology we saw in the last chapter at the Slough plant, or by changing land uses so as to reduce soil erosion on farms.

This is not only an issue for wild fish, however. The pollution of coastal waters also degrades the most important resource we have for fish and shellfish farming, namely that of clean water. And the consequences of marine pollution sometimes harm human health. Bacteria, viruses and blooms of toxic algae can affect the quality of seafood. It is estimated that between 1996 and 2000 seafood-borne illness affected about 116,000 people in the UK each year.

Then there is the employment side of the equation. In 1948 there were nearly 40,000 fishermen in the UK working more than 13,000 vessels and between them achieving a peak catch of about 1.2 million tonnes. In 2008 there were just over 10,000 people fishing from 6,575 vessels catching a lot less than half that. These bald statistics, indicating big changes in our fishing industry, of course hide a lot of complexity, including the massive impact of technology. One measure of that is how in the early twenty-first century a modern beam trawler is about 100 times more effective at catching plaice than was a sail-powered vessel operating a century before.

Whatever the technology though, demand for seafood remains huge. We eat on average 23.6 kilogrammes per person each year and in 2008 bought over 385,000 tonnes with a market value of about £2.73 billion. While much of our catch goes abroad we are ourselves becoming increasingly reliant on imports, which in 2006 were worth nearly £2 billion. Among the main species imported are those that were traditionally caught here, including cod and haddock.

As we come to rely more on imports, perhaps we should reflect on how past events led to our own fish resources becoming relatively more important for nutrition here. Back in the 1970s the international adoption of 200-mile territorial waters meant that British boats had less access to the fish in the productive seas around Iceland, Newfoundland and Norway, and instead fished much more on the grounds around our own coast.

In the 1960s the North Sea was the source for about 10 per cent of English cod landings, whereas during the 1980s it supplied more than 80 per cent. The better we manage our own waters, therefore, the better our chances of securing domestic supplies of seafood — surely a good idea in a world of fast-expanding demand, falling fish stocks and rising prices.

Fishing technology can be good

While ever more effective fish-catching technology, too many vessels (including the legal presence of international fleets under EU laws) and depleted stocks has led to the decline of many British fishing ports, some still thrive. One of them is Brixham in Devon.

Visiting the quayside fish market early one July morning I was impressed by the diversity, quantity and size of the catch. Boxes full of big brown lemon sole are lined up in neat rows, dozens of John Dory with their characteristic dark spot on the side sit smothered in ice. There are boxes of turbot, a few of scarlet-bodied red mullet and several of silvery slender hake. Pollack with their huge dark eyes stare out from another row. Line-caught cod with their intricately patterned green-and-white-speckled flanks sit alongside a good catch of graceful silver whiting. In a big box on its own is a blue shark.

An auctioneer conducts a noisy sale of the fish. As he goes from section to section a little coterie of bid-makers follow behind, among them local restaurants, traders from London and a buyer from Sainsbury's. That huge retail chain, in common with some others, has adopted a policy of only selling sustainably sourced fish and the supermarket man is buying all he can, especially the line-caught cod.

There's a lot to be said about the policies needed to keep fleets like that at Brixham in business – among them rules to prevent the discarding of unwanted fish, precautionary catch quotas, fishing gear that does less damage and the protection of spawning areas. There are big issues around how best to gather the data needed to manage stocks, and a lot of complexity (and controversy) linked with harnessing market power through eco-labels like the Marine Stewardship Council (MSC).

Above all this, however, is the attitude of the fishermen. Around Britain there are many emerging examples of how action by fishermen is leading to better fishing and stock recovery, including by trawlers operating out of Brixham. At a meeting jointly organised by SeaFish, the official body that promotes the sea fishing industry, and the Prince of Wales's International Sustainability Unit (to which I am an advisor), I met fishermen who've made changes to their gear, so as to protect the seabed and to reduce the catch of smaller and unwanted fish. Working with a grant from the Centre for Environment, Fisheries and Aquaculture Science, fishermen operating from Brixham have been provided with the resources needed to redesign their equipment and to work with fisheries scientists in refining it.

One change is aimed at reducing the damage caused by the heavy beam at the leading edge of the nets, which would otherwise smash into the seabed, in the process damaging the habitat of creatures that live there. By fitting wheels to the ends of the beam it is lifted off the seabed and rather than hitting the bottom it rolls along, gliding over the top.

Through other changes in their gear, discards (that is unwanted fish that would be thrown away) have been cut in half, and where non-target species are still being caught the fishermen are working with retailers to help market lesser-known kinds of fish to consumers. The effect has been beneficial for conservation and the marine environment, but it has also helped fishing businesses.

For a start, less fuel is needed to drag the nets along because there is less friction with the bottom of the sea; and due to less wear and tear the nets last longer, up to 14–16 months compared with 8–10 before. As a result of better catches achieved at lower costs, crew wages have gone up by about £200 per week. The fact that these kinds of gear changes are benefiting the fleet is reflected in how some 90 per cent of the twenty or so Brixham trawlers are now using them.

The growing realisation of how we could get more value from our seas through better methods led to the Brixham meeting proposing that a new training scheme be put in place for fishing-vessel skippers. The idea would be to equip the people running fishing businesses with the information and ideas they need on how to sustain the marine environment that pays them and feeds us.

England's coral garden

On top of action by fishermen, big differences can be made through more effectively planning how we use different areas of our marine environment. Work under way at Lyme Bay off the Dorset and Devon coast gives a sense of the possibilities.

Offshore here is a unique reef feature that has been described as 'England's coral garden'. Hosting species that include the pink sea fan and the sunset cup coral, the bay also supports fishing and recreation-based businesses. Following a government consultation exercise and a long and intense campaign run by the Devon Wildlife Trust, a 206-square-kilometre area was declared by the British government in July 2008 to protect the reef from damage by trawling and dredging, while at the same time aiming to protect the local economy.

Inside the newly designated area the use of mobile fishing gear, including that used to dredge the seabed for scallops, was stopped. Other fishing was still allowed, however, for example with lines and pots.

Professor Martin Attrill, Director of the Marine Institute at the University of Plymouth, leads a research programme that set out to understand the impact of the changed fishing practices in Lyme Bay and has seen first-hand what can happen when creative and joined-up approaches are employed. With five years of data gathered and analysed he is very clear as to the result. 'Without a doubt you get improvements,' he says. 'We've seen clear recovery in the reef organisms and we've shown changes in the fish and mobile invertebrate fauna. This is proof that if you stop dredging you improve the marine ecosystems in these areas.' And, he adds, some of the environmental benefits have expanded beyond the protected area, as reef fauna have spread onto the surrounding seabed.

Very importantly, the benefits for nature have been achieved at the same time as protecting local jobs and livelihoods. When it came to any negative economic effects Attrill reports that 'we couldn't find any of that at all.' While some fishermen have had to travel further to fish, and have loudly protested their displeasure at the loss of traditional access, they have continued to catch fish. At the same time, other sea-based businesses have found considerable *added* benefits, including companies offering dive trips, sport fishing and boat excursions.

There has also been a rise in the use of static fishing gear, however, which has brought new challenges. With the withdrawal of dredging there was a considerable increase in fishing with pots and lines for lobsters, crabs, rays, sole, cuttlefish and a massive 600 tonnes annually of whelk (mostly exported to Asia, where it is eaten for its perceived aphrodisiac qualities), to the point where the level of activity is unsustainable. The Blue Marine Foundation is working with fishing and conservation interests to find a solution.

Charles Clover is the organisation's founder and he told me how his plan is to get the new protected area to work better through helping the fishing businesses working in Lyme Bay to get more value from their products. 'We are attempting to get the fishermen more value from their catches by introducing a sustainable fishing regime.'

There are some surprisingly simple things that can help do this, for example installing ice machines at the fishing ports. By icing the catch at sea the shelf life of seafood is extended and can increase prices paid to the fishermen by up to a third. That in turn will mean they can make a living from smaller catches achieved with less time at sea.

Clover and his group, working in partnership with Attrill's team of scientists from the University of Plymouth, are seeking to build common cause among the fishing businesses, and he says that is working well. For a durable solution though he believes further regulations will be needed. While official bodies are reluctant to put in place more rules because of their fear of economic downsides, a change of mindset could instead see regulations adding economic value. 'If regulators looked at the opportunity in a more creative way, rather than only the costs, then we'd get much more benefit from this area of sea for the communities that live around it.'

Research that points in this direction has already produced some important results. It has for example been found that, contributing over £13 million, sea anglers have the highest total estimated expenditure per year in Lyme Bay. The boat charter and dive businesses that rely on marine wildlife have a combined turnover of more than £3.5 million per year. Researchers found that these kinds of businesses brought more into the local economy than commercial fishing; for example, landings from scallop dredging within Lyme Bay were valued at £1,848,577 in 2007, underlining the need to look at the bigger picture, rather than simply the 'costs' that come to any one economic sector.

Several factors working together have helped to achieve overall continuing progress at Lyme Bay. One is the considerable effort being put into partnership working between fishing and conservation interests. Another is high-quality research to establish exactly what's happening from both environmental and economic points of view. Most important, however, is the original official spatial planning decision from the government that set up the reserve in the first place. Once that was done all else became possible. While that work is not yet complete, with the need for further regulations evidently necessary, a good start has been made.

Unfortunately, the progress that was hoped for a few years ago in relation to the designation of a full network of 127 marine protected areas around Britain (never mind their effective regulation and administration) has not been forthcoming and the momentum that came with plans to do more to protect the UK's marine environment has stalled.

Citing insufficient science as the reason for a go-slow, Ministers backed off from their original plans and have at the time of writing established only 27 of the reserves they originally expected to protect under new laws. The Marine Conservation Society is one of several environment groups to express bitter disappointment at the decision, seeing the delay as a consequence of placing the benefit of scientific doubt with narrow existing economic interests, rather than in favour of nature.

Against this backdrop, however, the experience at Lyme Bay demonstrates what can be achieved through an integrated approach, not only in protecting natural systems, but also in reaping economic and social benefits in the process, including through helping businesses that are harmed by the status quo.

Back on the menu?

There has also been a positive effect arising from measures to reduce pressures on particular fish populations, including the iconic North Sea cod that during recent decades saw such a drastic reduction in stocks.

During the 1960s and 1970s cod catches from the North Sea were around 300,000 tonnes in most years, but by the 1980s they had declined to less than 200,000 tonnes. By the 1990s catches ranged from 100,000 to 150,000 tonnes. Worse was to come and during the 2000s the total allowable catch was set at less than 50,000 tonnes, with the nadir reached in 2007 when the overall catch was set below 20,000 tonnes.

One person who knows this history very well is Shetland fisherman John Goodlad. He's worked in the fishing industry for most of his life, comes from a fishing family and has been a champion for sustainable fishing in the Scottish industry for some years as chairman of various fisheries companies and organisations including Shetland Catch, Shetland Fish Products and the Scottish Pelagic Sustainability Group. I got to know John through working with him at the Prince of Wales's International Sustainability Unit, where he has brought an industry perspective to the complex task of charting pathways to sustainable catches.

On the reasons for the rapid decline of North Sea cod he told me that the answer can be summed up in one word.

'Technology'. He explained how the huge increase in our ability to catch fish has had such a profound impact. 'A fishing vessel in the 1950s could catch so much, but the same vessel in the 1970s and 1980s was able to catch five or six times more. The North Sea cod stock began to decline dramatically, and that led to many people saying it was finished.'

Having reached the low point, however, a package of different measures has helped to turn things around. These have included the setting of more science-based catch limits and the enforcement of these, a reduction in the size of the trawl fleet, limits on the number of days that vessels could fish for cod, avoiding areas where young cod congregated, and increased mesh sizes on nets so smaller fish could escape. Because of all this there are signs that the stock is beginning to recover, with the catch quota for 2014 set at 26,000 tonnes.

Goodlad is cautiously optimistic. 'When you talk to fishermen they all agree – whether they're Scottish, Danish, German, Dutch – they're all saying that they're seeing huge quantities of small cod and that's a really encouraging sign. That's the new-year classes coming through. They've survived. They haven't been caught and discarded as small fish. We are now in the early stages of what we hope will be long and sustained recovery.' That view is shared by fisheries scientists.

On top of the policy changes, the incentives and new fishing gear, Goodlad puts progress down to a change of heart among fishermen. 'They've had to undergo a huge cultural change,' he says, including from being hunters to businesses that are more like those based on the management of grazing animals on pasture. The equivalent of grasslands in this case is plankton growing in a healthy sea and the food webs of crabs, worms, small fish and all the rest that depend on it.

He explains to me that the main reason why they've been prepared to go on this journey of culture change is because of the huge prize that's at hand. 'The reward for fishermen is the knowledge that the North Sea used to yield three hundred thousand tonnes of cod every year. That's the reward that someday they can get back to landing that quantity of fish in a sustainable manner.' It's too early to declare a major success but Goodlad points out the importance of having big ambitions.

If there were to be a sustained increase in the cod population, then perhaps in the not too distant future we might reach the point where we can once more enjoy home-caught cod in the knowledge that stocks are in good health. We should be confident that this can be done not least because it has occurred elsewhere, notably in the Barents Sea off Norway, where the sustainable catch for cod is now set at one million tonnes per year. The Baltic cod is recovering too, with the population from which I caught that fish one summer evening from a Danish beach now growing again.

In the case of the Baltic cod there is a reminder as to why it is not simply a question of managing populations of particular species of fish that we need to get better at. It is also about the way we understand and manage the whole ocean system. Baltic fishers are reporting a huge increase in cod numbers, but the young fish growing toward a commercially valuable size are rather thin. They're hungry, because the food web in which they exist is presently not supplying enough smaller fish for them to eat. With this knowledge there is a suggestion that fishing for smaller species such as sprats needs to be scaled back so as to leave more for the cod to eat. It might also be a question of better understanding the migratory patterns of sprats, which move around in response to natural changes.

Certified fish

Once the science and ecology of fish stocks are better understood, and once there is a will among fishing businesses and governments to work together in implementing solutions, then recovery is certainly possible. One way to increase momentum in that direction is by certification of sustainable fish that appears in shops and restaurants.

Goodlad says that certification by bodies such as the MSC can help. 'It gives the fishing industry the international audit on best practice, and if you get certification, then that is somebody outside, somebody independent, saying that you're doing the right thing. North Sea haddock is now MSC-certified, North Sea cod isn't quite there, but with a management plan in place, the stock recovering, and all of these other measures in place I do not believe that it will be that long before we have an MSC accreditation for North Sea cod.'

Connecting consumers with positive progress via the market certainly seems logical in rewarding those using better methods and an ambitious programme is under way that over time will bring more fishing businesses into the MSC's certification scheme.

James Simpson speaks for the Marine Stewardship Council and explained the background to 'Project Inshore' and its efforts to unite fishermen and retailers in getting more of the British catch certified. 'Working with all sorts of people across the supply chain, including the likes of Sainsbury's and Marks and Spencer, the idea is to pre-assess every fishery in the inshore area to get a map of sustainability around the UK, highlighting where we need to work to identify which fisheries might be eligible to get MSC certification.'

While recognising the breadth of good work already going on and the potential for us to get more food from the seas around us, there are of course considerable challenges. One relates to how official policy-making that governs the management of fish stocks is largely determined through European Union negotiations.

The importance of this particular fact of life is reflected in how 75 per cent of the volume of fish caught in UK waters is captured by non-UK vessels, including by French, Danish and Dutch fleets. If both fish stocks and the health of the marine environment are to improve over time, then we'll need more than just British vessels adopting different methods.

Progress was made at the EU level during reforms to the Common Fisheries Policy agreed in 2013, but far more is needed. Moving forward will require consistent demand for higher standards from British (and other EU member states') ministers and industry players, including retailers and fishermen. Recent progress demonstrates how making the journey toward sustainable fisheries is, however, not only desirable but also feasible, if the will is there to do it, which it increasingly does seem to be – especially among many fishermen. The case of North Sea cod proves the point.

So it is that our food is supplied by nature. From the recycling of nutrients in soils to the work of pollinators and natural pest predators to the solar powered food webs that produce fish, healthy natural systems underpin our nutrition. When it comes to what nature provides, however, there's something of even more immediate and wide-ranging importance for our national wellbeing. Like most of our fish it too mostly originates from the sea, but instead of coming via trawlers and nets it arrives in clouds.

Manifesto

- **Implement** an integrated marine spatial planning strategy, so as to balance our seas, for food, energy, conservation and tourism. Fund a major new marine survey programme so that this can be science-based.

- **Fund industry bodies** to establish a new training scheme in sustainable fishing for all new skippers.

- **Adopt a leading position** in EU fisheries negotiations using examples of best practice among fishermen to shape future policies.

Heather burning on an upland grouse moor in Scotland – a practice that leads to higher water bills.

Chapter 4
Waterland

16.8 BILLION – CUBIC METRES OF WATER TAKEN FROM THE UK
ENVIRONMENT EACH YEAR

100 – PERCENTAGE OF THE UK ECONOMY DEPENDENT ON FRESHWATER

ONE QUARTER – THE PROPORTION OF WATER CATCHMENTS THAT HAVE A
'GOOD' ECOLOGICAL STATUS

Access to clean water at the turn of a tap is something we
take for granted in the UK. When we think about it at all it is
generally the bills we pay, rather than where the water comes
from, or how it gets to us. Behind the companies who charge us
for water, however, lies an even bigger utility. Nature.

Walking on Dartmoor in January I find myself literally amid
our water source. Here, some 500 metres above sea level, winter
rain pelts through thick grey mist. Deep-chilled, it descends in
flapping sheets across the moors. This is a maritime environment,
where the moisture arriving from the ocean meets the land. It's
wild, windy, west and wet. The highest areas of Dartmoor average about 2,000 millimetres of rainfall per year, but other uplands
receive double that. The Pennines, Lake District and the Scottish
and Welsh mountains are among the wettest places in Europe.
The eastern lowlands of the UK, by contrast, and especially
southeast England, are surprisingly dry. London is drier than
Rome and in some years Essex has less rain than Jerusalem.

Wet or dry, the amount of water is one of the fundamental factors that determine how places are. For Dartmoor it is a combination of high rainfall and cool temperatures that enables it to have a covering of peat. Accumulating beneath a living layer of *Sphagnum* mosses it has built up as a vast area of blanket bog. Aside from occasional rugged granite tors that punctuate the landscape, this vast sponge sprawls across the high windswept plateau, smothering the hills and smoothing contours. In some places this soggy organic cloak of un-rotted plant remains – built up over the last 8,500 years – is an impressive eight metres thick.

Standing in the expanse of bog that lies between the sources of the rivers Teign and East Dart, and looking in all directions to the softly rounded treeless horizon, it is easy to see why so many people have over the years come to regard this place as a desolate wasteland. As we'll see later, however, it is far from devoid of value. In fact, it is an economically essential piece of the UK's infrastructure. From agriculture to manufacturing and from energy to leisure, if there's no water, then there is no activity, no profits and no jobs. Everything comes to a dusty grinding halt. But while we all know this, it is remarkable how little this basic reality is reflected in our overall approach toward the natural systems that sustain secure supplies.

A hangable offence

Perhaps we are prone to take water so much for granted because in the UK it is everywhere: in the clouds, frequent rain, the ground beneath our feet, rivers, lakes and ponds. Many of our ecosystems are dependent on abundant water – bogs, fens, grazing marshes and reed-beds among them – and then there are the truly aquatic habitats – rivers, lakes and ponds – that are even more dependent, including on replenishment from subterranean water.

The fundamental importance of water has of course been recognised throughout history. A look at an Ordnance Survey map reveals how often the UK's hamlets, villages and towns were founded on rivers, streams and springs. It is also seen in the wide range of rules and regulations that have been set out to protect water. And these are not recent EU directives. As early as 1388, during the reign of Richard II, laws were enacted to make the dumping of animal waste, dung or litter into rivers illegal. Underlining people's appreciation back then of the importance of good quality water is the fact that hanging was a possible penalty.

Measures more recognisable as modern environmental laws include the River Pollution Prevention Act, passed in 1876 to protect rivers from pollution, and the River Boards Act of 1948 that introduced regulatory powers dealing with water supply and sewage treatment. Since then progressively stronger environmental legislation has been enacted to protect water quality and more latterly wetlands.

While today's UK law carries less draconian penalties than it did in medieval times it remains a vital factor in sustaining the public goods provided by water. So too does the EU Water Framework Directive, adopted in 2000. This requires member states to ensure that all surface waters are in 'good' condition, a standard determined by a range of water-quality and ecological indicators. It marked a big step in how we protect water. While for centuries we had focused on preventing pollution, the Directive was about the actual state of water bodies and whether they are fit to support life – fish, insects, birds, plants and the rest.

The deadline for achieving 'good' status is 2015, a date the UK is not going to meet, as currently less than one quarter of our waters meet EU standards (and recently the proportion has slightly declined). And things are set to become more challenging, as it is not just the water bodies that need to be managed to achieve 'good' status, it is their catchments as well – that is the areas of land where rain falls before it ends up in water bodies, wetlands or aquifers (underground water held in porous rocks).

Aquatic systems and wetlands are very sensitive to subtle changes in water supply and quality and these are often determined by decisions taken far away from the water bodies and wetlands. By making the condition of water bodies the legal objective, it will no longer be good enough to simply require factories and sewage works to reduce what they are releasing; everyone else affecting the water environment will need to act too, including farmers applying nitrates and phosphates to the land. All of which is easier said than done.

There are more than 389,000 kilometres of river, about 6,000 permanent lakes, nearly half a million ponds and 714 separate underground water bodies in the UK, and pretty much the entire land surface of the country makes up their catchments. On top of these are a number of wetland ecosystems, including 392,000 hectares of fenland, reed-bed, lowland raised bog and grazing marsh and nearly one million hectares of floodplain.

As the economy has expanded, water and wetland environments have become subject to massive and growing pressure and some have undergone drastic reduction. For example fens have shrunk from 310,000 to 26,000 hectares while nearly all the lowland raised bogs have gone. Then there are the effects of pollution. Our best rivers, those officially recognised for their wildlife value, are in a worse state than any other kind of similarly designated habitat in the country, mainly because of nutrient pollution. The principal culprit is phosphate. Some comes from sewage treatment plants but as we saw earlier a lot of phosphate gets into rivers via fields. When it does it tends to bring soil particles with it, causing more environmental damage.

This is why the pressures are especially intense on rivers that pass through urban areas (from which sewage is discharged) or intensively farmed land (where fertiliser and soil erosion are major issues). On top of this is the already discernible impact of climate change. The average water temperature in many British rivers has increased by between 1.5 and 3 degrees centigrade over the past two decades, and it could change further as the effects of climate change become more pronounced.

In addition to these pressures is the already massive and still growing demand for water. Each year in the UK some 16.8 billion cubic metres of water are taken from the environment. A lot of it comes out of the ground and will continue to do so. This is especially the case in the most densely populated areas, where demand is the highest and rainfall the lowest. In the southeast of England some 70 per cent of supply comes from porous rocks that store water, including greensands, sandstone, limestone and chalk. Especially important are the vast subterranean reservoirs in the chalk that runs beneath some of the most densely settled parts of England. A great many of our towns and cities rely on this layer of rock for their water.

Easy water and lost chalk streams

Those of us living in areas where the public supply comes from chalk aquifers sometimes use the term 'hard water' to describe its alkaline character and propensity to clog up kettles, dishwashers and showerheads. For the water companies though, it might conversely be called 'easy water'. It comes out of the ground pure and clean, filtered through the tight pores of the white rock, ready to drink, flush loos, wash cars and sprinkle lawns. Requiring minimal treatment, it's a cheap way of getting water to lots of people, especially when they live alongside.

The thick chalk deposit that runs beneath London and rises up either side of the capital extends across a wider swathe of southern Britain, stretching from the white cliffs of Dover in the east to the Dorset Downs in the west, and north into

Hertfordshire. There's also some chalk in other parts of the country, including East Anglia and Yorkshire.

Laid down by tiny organisms in the warm seas of the late Cretaceous period, the smooth rounded hills that remain today are not very high. Britain's most famous chalk landscape is the South Downs in Sussex and Hampshire, where the highest point is Butser Hill at 270 metres above sea level. Despite the modest elevation, chalk geology shapes the character of much of southern England and is responsible for some of our most treasured and iconic landscapes. With wooded flanks and thyme-fragranced springy grazed turf on the top, Iron Age forts commanding spectacular views, haunting calls of skylarks and stone curlews and little blue butterflies dancing between orchid spikes, England's chalk hills have for centuries been a magnet for walkers, naturalists, writers and archaeologists.

The filtered water that has in some cases spent decades percolating through rock is not only vital for public supply but maintains a steady flow and temperature to create the unusual circumstances that sustain chalk rivers. These in turn support characteristic communities of plants, insects and fish, most famously members of the salmon and trout family. There is much to be said about these very special ecosystems, but perhaps most remarkable is that England has 85 per cent of them – and that's not just 85 per cent of the EU total. That's 85 per cent of the *world* total. The few chalk rivers in the world that are not in England are in France and Belgium, making the streams that flow from the southern British chalk very special indeed.

But while chalk aquifers are good for our water supply, taking water from the chalk is often not good for chalk rivers. Take the case of the River Beane in Hertfordshire. Once a pure swift-flowing stream famous for excellent trout fishing, the Beane used to attract anglers and walkers from London, people swam in it and went on boating trips. No longer.

On the edge of the village of Aston, close to the eastern side of Stevenage, can be found the reason why. Here sit the squat windowless brick buildings of the Whitehall Pumping Station that each day extracts about 23 million litres of water from the chalk. So instead of flowing down the channel of the Beane as it once did, the water from the ground is now diverted so that it can instead flow from the taps and showers in the houses of the nearby town and villages. The effect on the river has been devastating, leading to one of the worst examples in the UK of environmental damage caused by taking too much ground-water. Although the impact has been unusually bad on the Beane, a similar situation prevails on many other chalk rivers.

Dr Charlie Bell, the Living Rivers Officer with Hertford-shire and Middlesex Wildlife Trust, told me about her work to conserve these remarkable rivers, several of which run through Hertfordshire and which are among the most threatened of all. 'The chalk is the source of nearly all of our drinking water,' she explained. 'The groundwater feeds the springs that feed the chalk rivers, but the groundwater is being depleted by taking it out for public supply. It's very low-cost water because it's clean and has been filtered through the chalk.'

The effect of taking millions of litres every day has been dev-astating for the river. 'It's little more than a dry ditch for much of its length.' It wasn't always like this of course. 'Some of the more elderly local people can remember swimming in it, and water voles in abundance, kingfishers, canoeing, but now long sections of the river are dry all year round.'

In addition to the impact on ecology, Charlie Bell says that another big downside is for recreation, for fishing and walking in particular. 'Hertfordshire has a big population and many towns, such as Stevenage, Welwyn and St Albans, have chalk rivers. The potential of a healthy river for recreation, and for engaging people with their environment, is huge.'

It would be entirely possible to help the Beane recover. However, Ofwat, the official body that regulates the private water companies in England and Wales, has in the past prohibited Affinity Water from spending money on developing alternatives to large-scale extraction from the chalk. Meanwhile, proposals for thousands of new houses, that would demand yet more water, have been given the go-ahead.

For the Beane the situation is made even worse by the fact that waste water from homes and businesses is not cleaned and returned to the river, as it is in many other places, but is instead transported in sewer pipes to the Rye Meads water treatment plant about 14 miles away in the valley of the river Lea. Years ago the Beane flowed into the Lea along a beautiful tree-lined channel, but water that would have once arrived there in a river teeming with life now makes the journey in sewage pipes.

So it is that our taps and toilets have literally replaced a once thriving chalk river. In the process many of the different values it provided to society and the economy have been lost in favour of just one – 'cheap' water. Considering the multiple benefits that a river like this could deliver for people, I'd say that this is neither an intelligent nor optimum outcome. Bell put this into context. 'There are about 180 chalk rivers in the world, which makes them rarer than giant pandas. They are incredibly beautiful and diverse ecosystems, or at least they should be.'

Progress is finally being made on the Beane, through the work of the Beane catchment management partnership, and in the years ahead it should begin to recover some of its former glory. But it is far from unique in suffering in this way. Many of England's beautiful chalk rivers, especially those in Buckinghamshire and Hertfordshire, are grossly over-abstracted. So are some of the very finest examples in Hampshire and Wiltshire, including the Kennet and Hampshire Avon.

Near to where I live in Cambridge is the rather distressing example of the Cherry Hinton Brook. This short chalk stream, less than three miles long, rises at a ridge of chalk to the south of the city and flows north to join with the River Cam just downstream from the main built-up area. In addition to issues of low flow, it is for much of its length also grossly polluted. In wet weather its lower reaches receive sediment, run-off from roads and other pollution to create a black plume that spews into the Cam where the two channels meet. That such a state should be tolerated for so special an ecological feature is troubling. For it to occur in a city that is proud of its global leadership in conservation and sustainability presents special irony.

As for the future, the requirements of the Water Framework Directive should help England's chalk rivers, assuming British Ministers don't negotiate get-out clauses as part of government's intended redefinition of the UK's relationship with the EU. Should the Directive survive then meeting its requirement of 'good' status will need a more integrated and long-term approach toward water.

For the densely populated south of England I have reached the conclusion that in addition to better management of water resources and more stringent pollution controls, new reservoirs must also be part of the solution. Topped up from rivers during periods of high flow in the winter, new water storage could help take pressure off sensitive ecology while improving water security for homes and businesses.

This would benefit not only chalk rivers but also many other kinds of streams and wetlands that are similarly afflicted by groundwater abstraction. Making the most of what nature does for us by replacing water taken from the ground with that from a reservoir is one option being considered in helping the Beane recover. In this instance the idea is to pipe supplies to Stevenage from Grafham Water in Cambridgeshire.

This might assist, but as demand continues to rise across much of southern England in line with population growth, an expanding economy and more development, the pressures on chalk rivers and other aquatic systems can only be expected to grow, especially as climate change creates more volatility in the water cycle. Reducing leakage from pipes, building more capacity for wastewater treatment and scaling up measures to improve water efficiency in homes, offices and industry will need to be part of the plan, but even if that succeeds new reservoirs will be needed.

We haven't gone down the reservoir-building route in recent decades in part because of local opposition. When it comes to threats to the local environment some people evidently regard construction of new reservoirs as comparable to a new motorway or opencast coalmine. As is demonstrated by many positive examples across the country, however, major water supply reservoirs can bring significant environmental improvements and social benefits that go far beyond water security and protection of rivers. New reservoirs have also been resisted on cost grounds and fears over the impact on consumer bills. This short-term economic perspective, however, ignores the potentially huge economic costs that would be caused by prolonged drought, especially in the southeast of England, where the 'headroom' between supply and demand decreases each year, exposing the economy to greater and greater risk.

It is not only the amount of water that determines the value of aquatic ecosystems to society, however.

Riverine resurrection: the Itchen

The Itchen is one of Britain's most beautiful and iconic rivers. A chalk stream, it rises on Hampshire's Cretaceous hills and threads south to join the sea at Southampton Water. Just upstream from the historic city of Winchester it flows across a floodplain called

Winnall Moors. Long before this exceptional river was officially recognised for its outstanding ecological value, this stretch was famed as a spiritual home of fly-fishing.

As fly-fishing became a hugely popular sport, it became tempting to fix rivers to match the anglers' (sometimes low) level of proficiency and high catch expectations. Stretches of the Itchen, in common with many chalk streams, were subjected to an over-zealous removal of bankside vegetation, clearance of aquatic plants that grow from the riverbed and heavy stocking with big trout brought in from fish farms that are larger and easier to catch than wily wild ones. All this helped make sure that visitors didn't spend too much of their day untangling line and went home with an easy-to-catch fish for supper. On some popular rivers the result is banksides mowed to neat lawns, canal-like channels, and sluggish flow supporting populations of dozy farmed fish. This is what happened to parts of the Itchen and its carrier streams.

Then in the 1990s the Hampshire and Isle of Wight Wildlife Trust took control of the Winnall Moors floodplain meadows, and in so doing became masters of the river, its many channels and the fishing rights at Abbots Barton that came with them. I walked there with Debbie Tann, the Chief Executive of the Wildlife Trust, and her colleague Martin de Retuerto, who has been leading work to improve the river and the floodplain of which it is an integral part. They told me about their plan to restore the river and meadows to a more natural state and to develop a viable wild game fishery.

At first the local fishing club (then leasing the fishing rights from the Wildlife Trust) was alarmed. They feared that re-naturalising the river would harm the fishing and be an insult to the river's famous history. The Trust became subject to vitriolic attacks. 'We were accused of being anti-fishing and of wanting to turn the place into a swamp,' Tann told me.

'We were attacked in *The Daily Telegraph* and even *The New York Times*. We have never been against fishing but when we suggested there might be a different way many within the fishing community didn't like what we had to say. We were confident though that our vision for the river and its floodplain was in the true spirit of the angling heritage.'

The Trust pressed on. Sections of river that had been dredged and cut into straight channels were returned to a more meandering course. Artificial weirs were removed and some areas made shallower with beds of gravel. This sped up the river flow, in turn keeping the stony areas where the trout and salmon lay their eggs clean and free from silt. Next to the gurgling gravel shallows there are emerald green patches of plants, some growing out of the water at the side of the channel, including patches of watercress. In places the cress has grown into dense clumps that are visibly holding back the flow, lifting the river level upstream.

On the riverbed is a distinctive community of plants that support the diversity of insects: water parsnip, starwort and water crowfoot among them. Pools have been put back in. Some of these deeper runs, where the water slows down, are home to big territorial trout. In one deep hollow I saw a dark torpedo-shaped fish. As we got closer it slowly sank out of sight, menacingly slipping into the gloom below. From its lair it can rise to pluck insects from the surface and ambush smaller fish that enter its patch. As I watched speckled brown trout and grayling, with their large sail-like dorsal fins, dart through swift clear water Retuerto told me how salmon and sea trout have returned in some numbers, migrating up from the sea. The river is no longer stocked with farmed fish and is instead inhabited by wild ones. 'Because of the habitat improvement and reconnection with the floodplain, spawning and juvenile numbers have increased.'

The river has been transformed, and the water meadows too are beginning to look more like they did a century ago. The grassy areas that flank the Itchen were once integral to the local livestock industry, with pasture production boosted in the early spring by deliberately flooding the land. Inundation from the chalk river with water up to ten degrees warmer than the air thawed the frozen ground, thereby kick-starting the growing season earlier than would otherwise have been the case. In addition to warming the soil the river brought nutrients that also boosted grass growth.

The Wildlife Trust's own British White cattle graze the meadows and help maintain a vibrant mix of wildflowers, including marsh marigolds and orchids, while producing excellent meat. It's believed the Winnall Moors meadows stopped being regularly flooded in the 1930s – quite late compared with most others, where traditional practices ended around the time of the Great War. There's no one left with direct knowledge of how it all worked, however, and Tann told me how she and her team are having to learn methods that began to die with the men who went to fight in the trenches.

This kind of ecological restoration can clearly bring huge benefits for people's enjoyment of the water environment, and perhaps one day the Beane will benefit from similar treatment. But it's not just for recreation and nature that we might see a strong case for restoring degraded aquatic environments. Many benefits can also be gained for the security of public water supply – in catching rain, storing it and slowly releasing it in a pure state. One place where that connection between taps and the natural world can be graphically appreciated is back up on Dartmoor.

Upstream thinking

Dartmoor's rugged scenery is recognised for its outstanding landscape and wildlife but less often regarded as an essential part of national infrastructure. Yet that is exactly what it is.

Heading up by road from Exeter, crossing the River Teign, passing by the little town of Chagford, on a winding road to the Fernworthy Reservoir, and then on foot up through a mossy old plantation of Sitka spruce, you reach the edge of the open moorland. A beige expanse of purple moor grass lies ahead. Stretching up toward two Neolithic stone circles, onward to the untidy grey pile of granite called Sittaford Tor, and the land becomes wet enough for blanket bog to form.

The expanse of bog holds a vast quantity of clean fresh water and this living sponge releases it slowly into the rivers that rise up here, including some of the region's most beautiful and famous. But there is a problem. For too long regarded as a valueless wasteland, the high bogs of Dartmoor have suffered a catalogue of abuse. To calibrate their weapons for long-distance encounters at sea, Royal Navy battleships once used the high moor for target practice, firing massive fifteen-inch-diameter shells from vessels in Plymouth Sound. Blasting the living skin of the peat to smithereens, the heavy sea guns were echoed by smaller artillery rounds fired over army ranges up until the early 1990s. Less dramatic but more widespread have been the effects of heavy grazing and burning. Across broad sweeps of the moor the long-term disintegration of the bog is the result.

In some areas the smooth peat duvet is crumpled and creviced. From these damaged areas streams run out of the bog, taking particles and dissolved peat with them. Big chunks sometimes float off into the rivers below. When the peat gets into rivers and from there to water treatment plants, a lot of expensive technology is needed to clean it out again before it is piped to homes and businesses.

Andy Guy works with Natural England, the government's official nature conservation body, where he is an expert advisor on the Dartmoor Commons. He can read the land like a book. He showed me an area of degrading peat and explained how the process accelerates as water runs across damaged areas, cutting further back into the peat mass. 'Once that process is initiated the only way that you can stop it is to physically intervene. If you don't then the erosion gullies carry on cutting away.' There are places on Dartmoor where the peat has entirely gone.

Blanket bog degradation on Dartmoor is generally started by a combination of heavy grazing and burning. Guy points to a recently affected patch of bog. 'An area here has been burnt into over the years by fires. Where the peat mass starts to stand proud and becomes drier you can see that if a fire comes across here it will start to eat into the peat and accelerate erosion.'

The consequences are costly. Telling me why in the rain and wind on Dartmoor was Professor Richard Brazier, a hydrologist based at the Geography Department at the University of Exeter. He is a soil specialist and among other things leads on research looking at the restoration of bogs. 'If we see bare peat in contact with water there's going to be dissolution of the carbon, it gets into the water as humic and fulvic acids and so you get the export of carbon and associated colour moving downstream. Obviously the water company is interested in that as they've got to clean it up.' The acid also impacts on fishing rivers and the peat degradation has caused rare breeding birds, including Golden Plover and Dunlin, to suffer sharp declines.

As well as being an intricate ecosystem comprised of mosses, grasses, heathers, rushes and herbaceous plants, underlain by a mass of granite, with the whole lot running on water, the moor is what is increasingly referred to as 'green infrastructure'. For not only is this high raw place supporting wildlife and providing amazing recreational opportunities, it also helps to sustain the region's largest cities of Exeter and Plymouth. Like every other commercial centre on Earth these two regional economic hubs need copious supplies of clean fresh water to keep them going. The high hills that are the backdrop to their urban sky-lines, cafes, offices, workshops and shops are part of the answer.

So it is that this evocative and iconic upland is among many other things a valuable water supply asset. Intercepting the abundant clouds and masses of moist air that roll in from the Atlantic Ocean, the ancient plug of volcanic rock looming over the South Devon landscape is a massive natural reservoir and water purification system.

The fact that the moors help catch, store and supply pure water is no longer so taken so much for granted as it has been in the past. South West Water, one of the UK's private water supply and sewage treatment companies, has realised how the moor is helping to cap the company's costs, not least because the upland mires and grasslands, when intact, prevent particles and brown colouration getting into the water, thereby avoiding the need for so much equipment and chemicals to get it out again. By work-ing with the moor and its natural systems, the company is finding that the costs of supplying freshwater can be drastically cut.

As well as understanding how the moor works, Andy Guy is very much focused on the complex task of getting all the differ-ent interests on board with bog restoration, especially those who farm. 'This is the first time that anyone has tried to do blanket bog restoration on Common Land,' he says. 'We've had blanket bog restorations on private land in the UK for decades now, but that's relatively simple.'

A similarly complex situation faces South West Water in attempts to restore blanket bogs on Exmoor. Up there during the nineteenth century, on the top of the sandstone massif that is such a prominent feature of the North Devon and West Som-erset countryside, dense networks of drainage ditches were cut so as to get water off the land to dry it out for grazing. Although the land proved poor for rearing animals, the drainage ditches did succeed in shifting water off the land.

Brazier told me about the effect of the ditches and about recent attempts to reverse the damage they've caused. 'Some of those features are a hundred and fifty years old but they still shoot water off incredibly fast and you get these flash floods. There's more than two hundred square kilometres of Exmoor National Park and a large part of that is shallow blanket bog that is damaged in this way, so the intensive work undertaken since 2009 has been to block those ditches. Every five metres or so there's a block installed.' Using natural materials from the local area to block the drains, including straw bales, wood and stray lumps of peat, the huge water-storage capacity of Exmoor is being improved. Brazier told me how the work has been a success, including improved water quality. Also impressive was the speed of recovery. 'When you return a year or two later most people can't even tell where the restoration has occurred.'

By working with local people who use the land up there the company has led the process of improving the water retention of the uplands. This in turn is helping rivers to run more steadily and that has delivered a commercial benefit for South West Water because the company now needs to use less of its reservoir water to top up rivers during periods of low flow. That in turn is saving money and keeping down bills. Research published in early 2014 confirmed the benefits, with the contamination that causes discolouration of water found to be reduced by half, thereby cutting the costs of getting it out later on. This drop in dissolved carbon also means that more of it is staying in the bog, thereby adding a climate-change advantage too.

South West Water is also finding benefits in working with people who use the land in the lowlands. In this they have found a strong convergence with the work of the Devon Wildlife Trust that is seeking to conserve the so-called Culm grasslands – special areas of semi-natural grazing meadows that host an unusually diverse mix of animals and plants.

The total Culm area, where such grasslands *could* occur, covers about 2,800 square kilometres of Devon countryside. Today, however, only about 1.4 per cent of the original grassland remains. The rest has mostly been turned into 'improved' pasture. 'Improved' fields are for the most part heavily fertilised, species-poor and have compacted soils. Pete Burgess is a driving force behind Devon Wildlife Trust's efforts to maintain and restore the Culm. He told me how 'The rate of change in the last fifty years to the Culm and the surrounding area is astonishing. We've got 1947 aerial photographs and you just see these really intact landscapes.'

Although what's left is very fragmented, some of the remnants are fortunately concentrated in particular key areas. One is around the headwaters of the rivers Tamar and Exe, where Trust staff advise farmers on how Culm areas offer benefits through absorbing and retaining more water than managed pasture. This is because their soils are less compacted and because they contain more organic matter. Protecting the Culm grasslands can thus help rivers to flow more evenly while conserving amazing wildlife. While efforts to conserve the Culm might not be as visible as tree planting or hedgerow restoration, they can be even more dramatic in improving the state of the rural environment.

Efforts by South-West Water to improve catchments arises from an approach that they have called 'Upstream Thinking' (literally looking upstream to solve problems before they demand expensive solutions downstream). It's predicated on the kind of calculation that suggests how over a thirty-year period a £10m investment upstream could save £650 million of costs that would otherwise be incurred.

The more integrated approach championed by Upstream Thinking has enabled South West Water to work with farmers, conservationists and scientists in ways that enable better cost management as well as a range of other benefits. South West

Water estimates that by 2020 the company will be applying this kind of integrated approach to about 8,200 hectares.

Dylan Bright is head of sustainability for the Upstream Thinking initiative. He says a shift in mindset was important in making the progress that has been achieved. 'South West Water got a real interest at fixing a problem at source and not only for philosophical or virtuous reasons, but because of the strong economic case for doing so. It really is a win–win–win, and the beneficiary isn't just the farmer with his enlightened self-interest, it's society for all these other reasons – wildlife, cheap water, five hundred per cent increase in fish stocks, and so on.'

We all pay for grouse shooting

Whereas protecting Dartmoor's boggy 'wastelands' and conserving the insect-rich Culm grasslands once seemed like a cost, today they look like sound investments. And South West Water is not alone in its approach. Similar schemes have been adopted in various parts of the UK, including the wild Garron Plateau in County Antrim, Northern Ireland.

On a visit there, and to the Dungonnell Water Treatment Works, I found myself staring at what looked like a giant chocolate brownie. But while the colour and crusty form were reminiscent of the confection, the smell wasn't. A peaty aroma instead filled the air, and unlike cake on a plate this brown material was gently quivering. The pulsating movement was down to the fact that the crusty substance floated on bubbling water. In a rectangular concrete tank about five metres long, three wide and two deep, tiny air bubbles were being introduced from beneath. I was looking at water-purification equipment and was about to find out why such a bizarre-looking and elaborate process was needed.

The Dungonnell works are fed from a small reservoir that in turn is filled from a 2,000-hectare catchment covered with blanket bog (part of the largest such system in in Northern Ireland). Like many other similar ecosystems across the British Isles, the bog was not in best shape. Drainage ditches had been hacked across the bog and it was heavily grazed by sheep. Colour released from the bog and into the water presents a health hazard and has to be cleaned in accordance with official standards before being supplied to the public – hence the crusty peat topping on the 'brownie tank'. Crystallised out of the water using aluminium sulphate, the colour forms into particles which are forced up into that floating crust, which is skimmed off before being sent to a sewage treatment works. Nice clear water is produced, but at considerably higher cost than if it had arrived in the works with less colour in it to start with.

With a daily output of around ten million litres of clean water the treatment tanks stripping out colour require a lot of expensive chemicals to keep them going. The cost of that is inevitably passed on to consumers, 30,000 of whom around Ballymena are supplied from the Dungonnel plant.

Northern Ireland Water thought restoring the blanket bog might help reduce the colouring – and the costs – and in 2013 a partnership was forged with RSPB and the Environment Agency to do just that. Contractors were hired to block drains while agreements with farmers led to the number of sheep being halved. Positive results didn't take long. Walking on the bog one late summer afternoon, under grey skies and a mist of cool drizzly rain, I was surprised to see how quickly the vegetation on those damaged areas was recovering. A deep roadside culvert that had been dammed every few metres along its length only a year before was starting to look like a natural bog, its dams covered in vegetation, with little pools above, looking like those on the open moorland.

The result is not only a rapidly recovering bog that is helping to render water supply more cost-efficient, but also the conservation of the wildlife that lives up there, including Red Grouse, Hen Harriers and Merlins. Other rare wildlife that is benefiting from habitat restoration includes the Marsh Saxifrage, the intense yellow blooms of which grace the bog during its summer flowering time. This is the only place in Northern Ireland where this very rare plant is found, and efforts to clean up the water should improve its prospects too.

Roy Taylor is Catchment Manager with Northern Ireland Water. He explained the thinking behind the shift of emphasis from chemicals to the land. 'We're trying to restore the wetland to bring back the flora and fauna but also to get cleaner water into the treatment works. We can see the vegetation coming back very quickly. The mosses are returning in just a few months. It's difficult to predict savings but from what we are seeing among other companies the payback should be very quick. We've spent twenty thousand pounds on this and I know that won't buy you a lot of aluminium sulphate.'

It's another great example of joined-up thinking, addressing several challenges at once and in the process cutting costs. It has been possible to do this in part because Northern Ireland Water owns the land where the bog restoration is taking place.

On the Pennine hills of northern England, Yorkshire Water enjoys a similar level of influence over at least some areas of water catchment, and in those places the company is able to negotiate agreements with tenant farmers. Andrew Walker is the company's Catchment Manager and works with the farming community. 'I want to work with them and understand their business, and the ways in which we can manage land differently so that it delivers a benefit to them and a benefit to me.' His approach has led for example to agreements with farmers on the grazing of heavy cattle that break up thick grass, thereby enabling other plant species to thrive. This improves the quality of the catchment for water, while also bringing wildlife benefits.

Farmers working on land owned by the company are one thing, estates controlled by commercial grouse shooting interests quite another. Blanket bogs covering land managed for this activity are often deliberately burnt to encourage more birds, and among other things this causes a lot of colour to come off the hills in the water they shed.

Walker says that engineering solutions would require more investment in more expensive treatment works and water treatment processes. 'We'll always be able to treat the water, it's just that it's very expensive.' Another option is to look at ways of influencing land management, and specifically measures that will help bogs recover.

Some estates are reluctant to join in with blanket bog recovery, however, as they regard such work as hostile to the aim of increasing grouse numbers. Yorkshire Water is among those looking for ways to do both, and is convinced it can be done. 'In the uplands we do believe that dry, degraded moors are not as potentially productive as they could be, even for grouse. They're certainly not delivering clean water,' he says.

One lever society has for influencing how the land is managed is the vast amount of money paid from our taxes to upland shooting estates. From January 2015 the money handed out to moorland owners nearly doubled from £30 per hectare per year to £56. Perhaps if payments were linked with measures to improve upland habitats the public interest would be better served. Under current arrangements I fear more will be spent on new gamekeepers than it will be on bog restoration.

In the meantime the sport of a privileged few will continue to be subsidised not only through our taxes but also through the water bills of the many, who pay for the colour to be removed. This is economically irrational and patently unfair. When politicians bemoan the high price of environmental standards in our water bills, it strikes me that the ones among them who shoot grouse don't necessarily speak for those on low incomes who get water bills delivered to their rented flats.

Farming for water

For Northern Ireland Water, Yorkshire Water and most other water supply companies there are also major water-quality challenges in the lowlands. Around the Evington Treatment Works near York the company is trying to get across a different point to those who manage the land. It is about the loss of farmers' core asset – their soil.

Each year the treatment works removes about ten thousand tonnes of soil from the water it abstracts from rivers. That is not only a cost to the water company, but also to the farmers. Without it their businesses are of course nowhere. In this case the soil is settled out, dried and sent to landfill. Aside from the costs of this for the company's customers, the loss of our main food-growing medium is, as we saw earlier, a major issue.

Walker sums it up thus: 'Water quality and water management is all about soil management, keeping soils healthy and where they belong. Where there is a peat soil in the Yorkshire uplands and an agricultural soil in the lowlands, keep it there, keep it healthy, keep it full of organic matter, and then it will deliver that percolating filtration, it will deliver the crops that we need. It doesn't belong in our treatment works.' By working with, and offering advice to, the people who grow food, not only has Yorkshire Water been able to cut the level of soil arriving in its water treatment plants but also pesticides. The farmers are benefiting too, by saving money through using fewer chemicals and hanging on to more of their soil.

Another company that is working with farmers to keep rivers healthy is Wessex Water. Nutrients are top of their list, not least because of the environmental damage being caused by the build-up of nitrates and phosphates in the waters of Poole Harbour. This shallow bay just to the west of Bournemouth in Dorset is one of the richest natural features on the South coast of England. It is among other things a haven for wildlife, a major tourism asset and a commercially important source of shellfish. Connected to the sea by a narrow tidal channel, it is also connected to the land by a number of rivers, including two beautiful Dorset chalk streams: the Frome and Piddle.

These two waters flow through rural catchments that although picturesque are subject to serious environmental pressures. These are reflected in how the groundwater, many of the rivers' tributaries and the rivers themselves are failing to meet a number of standards, including the level of nitrate allowed in drinking water supply.

Wessex Water realised that if these rivers and Poole Harbour were to meet the target of 'good' status required by the EU Water Framework Directive, then the company needed to work with farmers rather than simply relying on yet more expensive pollution reduction equipment, since far more of the problem is down to farming than to inadequate sewage facilities. In the case of nitrogen an estimated 2,400 tonnes arrives in Poole Harbour each year, with around 80 per cent of that coming from farming and most of the rest from sewage works. Between two thirds and three quarters of the phosphate pollution comes from farming.

The agricultural nitrogen mainly gets into the Harbour via the groundwater, in some cases thirty years after it was applied to fields. The phosphate takes this route too, as well as being washed off fields attached to tiny soil particles. When these nutrients get into the rivers and then the Harbour they cause algal blooms followed by depressed oxygen levels and then a range of other environmental problems, including fish kills.

Wessex decided to take a long-term and integrated approach based largely on engagement with farmers. A team of Wessex advisors now offers a free advice service on soil and water quality, optimal fertiliser application and agricultural management plans, as well as assisting farmers in complying with conditions that must be met to access different funding schemes. The company is also innovating with alternatives to more technology in sewage works, including the use of reed-beds to cut phosphate releases. When phosphate-rich water is run through reeds the plants take up the nutrient as they grow, cleaning the water as it passes through their dense root-mat. It's a low-cost and low-energy method, and although it needs more land than an engineering-based solution it brings many other benefits, including the creation of wildlife habitat.

The steps taken by Wessex Water seem to be working and delivering a wide range of benefits. Farmers taking action to cut phosphate pollution are now losing less soil than before and in some areas of the Frome and Piddle the release of nitrogen is going down with dramatic effect.

Paying three times

These examples of positive action from leading water companies demonstrate the power of so called 'catchment management'. This phrase describes the idea of looking at the whole system that shapes water quality and availability, compared with 'end of pipe' engineering approaches involving concrete and chemicals that deal with particular problems, such as specific pollution sources. Unlike concrete, however, catchment approaches can be flexible and moveable – and considerably cheaper.

One review that I was personally involved with, and that was commissioned by three water companies, estimated that catchment-based approaches can generally achieve the same outcomes as engineering-led ones at about half the cost. When it comes to the wider societal benefits achieved from the investments the disparity is much bigger. Industry appraisals of different projects and programmes give a range of estimated benefit-to-cost ratios from 1:4 to 1:63 (that is in the extreme positive case generating value worth 63 times the investment). This huge value multiplier was calculated in part through looking at the benefits that were not the direct responsibility of the water company, for example the conservation of wildlife, protection of the marine environment and flood protection (all of which fall outside their core business remit, but which are of course important to society). Taking this wider view – which has the virtue of being based on reality – reveals how we are losing value and wasting money on engineering works, when we could get more from looking at the wider landscape.

Despite the evident opportunity to achieve what water companies must do at less cost, it is disappointing to know that under 1 per cent of planned water company expenditure is currently directed toward catchments. In other words the vast majority of resources is still aimed at more traditional engineering solutions. This is especially worrying considering how billions of pounds-worth of investment will soon be mobilised by British water companies to meet the requirements of the Water Framework Directive. Perhaps we'd make faster progress at less cost if there was more appreciation among policy-makers and regulatory agencies as to the unfortunate situation we find ourselves in, whereby in cleaning up places like Poole Harbour and the water running off the hills of Dartmoor and grouse estates, British citizens are paying three times. We pay once in tax-funded subsidies to farmers and landowners, we pay again in our water bills and a third time in picking up the tab for repairing different kinds of environmental damage, including the effects of flooding.

Surely more could be done by government and its official agencies to combine different rules, regulations and budgets so that the range of benefits we need from clean water can *all* be achieved as cheaply as possible? Part of the reason for the present fragmented approach arises from how some things come above all others – for example the illusion of 'cheap' food – and the mistaken view that protecting the water environment is somehow too expensive.

South West Water's Dylan Bright again: 'On the last round of Upstream Thinking the cost will be 65p on the water bill of each household per year for twenty years. It's really small beer, and if you have the chance to explain it to a village hall, all the people would agree that if that's what we're doing with it they'd be delighted.' I suspect they'd also be pleased to hear that the kinds of investments he told me about can also help in reducing flood risk – which is where our round-Britain tour turns next.

Manifesto

- **Require the regulatory bodies** responsible for water pricing and the environment (Ofwat and the Environment Agency in England) to work together and with water companies in devising integrated strategies for the improvement of the water environment, while keeping down consumer bills. Add a new duty on these bodies to control costs for future consumers.

- **Recommit to UK leadership** in meeting the requirements of the Water Framework Directive, recognising the multiple benefits that come with a healthy water environment, including those from the renaturalisation of rivers and wetlands.

- **Work with the water industry** to take forward plans for new reservoirs in water-stressed areas, especially in the south and east of England, incorporating leading design to achieve benefits for water supply, conservation, recreation and tourism.

One of the Loch Dubh beavers working on structures to hold back water. Our own efforts to do that don't always work, as demonstrated by these swans on the A149 at Cley, Norfolk, in December 2013.

Chapter 5
Taming the flood

42 PER CENT – PROPORTION OF FLOODPLAINS IN ENGLAND AND WALES PHYSICALLY SEPARATED FROM THEIR RIVERS

24 – NUMBER OF EUROPEAN COUNTRIES THAT HAVE REINTRODUCED BEAVERS SINCE THE 1920S

AROUND 30,000 – NUMBER OF PREMATURE DEATHS CAUSED ACROSS EUROPE BY THE 2003 HEATWAVE

Late January 2014 and the fields on Tealham Moor are under deep floodwater. Two weeks before Christmas the patchwork of pasture dissected by watery ditches known locally as rhynes was still green and grassy. Now it was like a vast lake. Three families of whooper swans visiting Britain for the winter have stopped by; six adults accompanied by six youngsters reared on Icelandic tundra the summer before. The magnificent birds look very much at home on land that has returned to its former state as an expansive wetland. But while the birds were literally in their element, many local people weren't. These flat low lands are a part of the Somerset Levels, and like much of the rest of this unique landscape are prone to flooding.

This is not surprising considering how the fields are near to sea level and surrounded by the Mendip and Polden Hills, which being in the west of England receive a lot of rain. During the winter of 2013-14, however, it was exceptional. Starting in December, depression after depression wheeled in from the

Atlantic, bringing repeated deluges right across the country, especially to the southwest of Britain, including to these flat damp lands and the surrounding high ground that funnels water to them.

Although some parts of the Levels regularly flood, the inundation was by modern standards out of the ordinary. At its peak more than 60 square kilometres were submerged beneath more than 60 million cubic metres of water, in places more than two metres deep. Roads were blocked, communities cut off, houses ruined and farming devastated. Sedgemoor District Council declared a 'major incident', thereby triggering an intensified effort from official agencies in providing assistance to the thousands of people and businesses affected. Prime Minister David Cameron, as we noted, promised a blank cheque.

The flooding of 2014, although dramatic, should, however, be put into perspective. That perhaps starts with the fact that some 635 square kilometres of Somerset lie below sea level, and therefore are likely to attract water. Before the drainage of the Levels' tidal marshes for farming began about 1,000 years ago, and before various modern flood defences were built, the place was subject to regular and extensive flooding. Even as late as 1919, after major drainage works had been completed, some 280 square kilometres of the Levels and Moors remained under water – more than four times the area flooded in 2014.

The history of human settlement in this area of wetland is in part reflected in local place names. Muchelney, cut off for weeks by floodwaters in early 2014, means 'Big Island' in Old English. Many other place names similarly indicate a wet and watery past – Godney, Pitney, Thorney, Bradney and many others, that through the 'ey' or 'iey' suffix that means 'island', say something as to the kinds of conditions that once prevailed here. It also says something important about where settlements used to be located – that is, on the islands and not in the wetlands.

Then there is the name of the county of Somerset itself. This appears to be derived from how prehistoric people grazed their animals on the rich low-lying grasslands and marshes during the summer, leading the place to be known in ancient times as 'Sumersata' – land of the summer people. It was too wet to use for winter grazing, so people didn't do it. Ancestral wisdom is also evident in many of the pictures of flooding seen on British TV screens in recent times. Many of the villages and towns surrounded by floodwater show ancient churches and houses sitting just above the muddy deluge, whereas the houses and buildings in the floodwater are modern.

The reaction to the inundation of the Levels in early 2014 also said a lot about our modern mindset toward managing flood risk. The default position of many politicians, parroted by many commentators, was to demand bigger defences and to dig rivers deeper, so that they would take water away more quickly. But might there be a case for a different and more integrated approach, one that might work better, and which could bring other benefits for less money? It seems there could be, but it requires us to look beyond the places that are flooding, and into the areas where the rain first falls.

Head for the hills

Five miles north of the mid-Wales town of Llanidloes a footpath heads out through a forestry plantation and up toward a group of peaks known as Pumlumon, the highest part of the Cambrian Mountains. The track follows a tumbling rocky stream, flanked by moss-clad boulders and in shady moist hollows patches of liverworts grow, there are clumps of blaeberry, a profusion of herbs and young deciduous trees just coming into leaf. Between them they create a riot of vivid greens that seem to scream in delight for the spring sunshine, mild temperature and abundant moisture that gives them such energetic life. Climbing higher and the stream is narrower until emerging from the top of the forestry and out onto the open moorland. I can easily jump across it. Yet that little stream is none other than Britain's largest river – The Severn.

About twenty minutes further on and the stream has diminished to a tiny trickle that emerges from a mossy pool: it's the ultimate source. From this tiny water, however, momentum quickly accumulates, for with average annual precipitation at about 3,000 millimetres (118 inches), Pumlumon is one of the rainiest parts of our country, and when a lot of it falls at once, the giant awakes.

Starting from these open uplands the river falls into valleys where fragments of the once extensive cloak of Atlantic oak rainforests still cling to some of the hillsides, and where wet woodlands with alder and willow still linger in some of the floodplains. Lower down the channel becomes broader and slower, flowing through farmland and then some of England's most historic cities – Shrewsbury, Tewkesbury and Gloucester – and on to the vast estuary where the Severn meets the Bristol Channel and then open sea. What happens along this long length of river is in large part determined by events in its headwaters. That's why at several points by the side of that tumbling upland stream concrete structures are built around it – flow-monitoring stations – that automatically gather data and send it for analysis so as to anticipate potentially devastating impacts downstream. Heavy rain falling up here gets quickly concentrated into a torrent, and measuring the changing state of the river can buy precious hours of warning for those at flood risk lower down.

In common with the Somerset Levels the valley of the Severn has during recent years experienced serious flooding. When analysing why it has been so bad it is important to look not only at the record-breaking rainfall figures and the state of various defences built along stretches of the 354 kilometre-long course of the river, but also to the more than 11,000 square kilometres of land that catch the water that fills it up.

A good place to start is where the river does. Befitting a river of such standing, there is a monument marking the point where it begins its journey to the sea. Just a few metres from the stone column that marks the spot there are clues as to why the flooding has been getting worse.

The blanket bogs that clothe the Cambrian Mountains have, as we saw on Dartmoor, been badly damaged, through grazing and drainage in particular. As a result, the land can hold less water, and instead of pausing in the hills, it heads straight off and down toward the sea, propelled by the relentless force of gravity. The more damaged the land, the quicker it leaves.

Further downstream too, on those green hills covered with sheep that are so characteristic of mid-Wales, the land is also discharging water more quickly than it used to. Compacted by hooves and made bare of trees so as to expand the area available for animals, the hills shed both water and soil on a grand scale.

This is no accident. In common with other upland areas of the UK, this part of Wales was 'improved' so as to enable more animals to be reared. Government payments encouraged large areas of moorland to be converted to permanent pasture through drainage, ploughing and the application of lime to counteract the acidity of the peaty soils. 'Improved' grass varieties were seeded and money offered to increase the number of animals. All this led to a sixfold increase in sheep numbers between the 1960s and 1990s.

While this might at one time have seemed sensible, the consequences of this 'improvement' downstream should cause us to pause before advocating more of the same in the cause of 'cheap' food. Events during the summer of 2007 show why. It was one of the wettest on record. Heavy rainfall at the end of

June led to flooding in parts of the Severn valley, but during July it got worse. On the 20th, two month's worth of average rain fell in just 14 hours, resulting in devastating floods.

The TV news that night – and for many days – showed the town of Tewkesbury turned into an island by a vast expanse of water delivered not only from the Severn but also the Avon that joins the bigger river there. Dramatic pictures of the ancient abbey surrounded by brown water were carried in news reports that went around the world.

Despite the distribution of 25,000 sandbags, over 1,800 houses and 300 businesses were flooded. For the first time in its 100-year history the Mythe Water Treatment Works flooded, resulting in the loss of mains water for 140,000 homes over a period of two weeks. Tankers and bottled drinking water had to be shipped in. An electricity substation only narrowly avoided inundation through the intervention of the emergency services.

Better flood defences could make a difference in defending vulnerable towns like this one, but there is huge potential to reap benefits by looking upstream as well, to where the water comes from in the first place. The Wildlife Trusts in Wales are seeking to realise this through a partnership with the Welsh government and others in their Pumlumon Project. It is one of the largest initiatives of its kind in Europe, embracing some 40,000 hectares of the northern Cambrian Mountains.

As we saw on Dartmoor and Exmoor in the last chapter, degrading blanket bog holds less water than a healthy equivalent. Rainwater runs off, increasing the chances of flooding further downstream in the valleys. Restoring the sponge-like quality of the bogs can thus help protect land and property downstream. Achieved through a combination of changed grazing practices to encourage more diverse vegetation, habitat recreation and the planting of trees and hedges, the project is effectively rebuilding the nation's infrastructure.

The restoration of more natural habitats is also helping declining wildlife species, and among other things that will support more tourism interest in the area. All this is being done at the same time as bringing community benefits, for example through helping local farm businesses diversify.

Clive Faulkner is Trust Manager at the Montgomeryshire Wildlife Trust and has been closely involved with the Trust's work on Pumlumon for more than twenty years. His job was initially concerned with reversing the decline of wildlife in the area. 'The more we looked at it, the more we discovered a whole range of problems – socio-economic, political, environmental – and it was at that point we had a choice. We either ran away with our hands in the air, or we sat down and thought it through. We did sit down and realised that they're all expressions of the same issue. That was the "Eureka!" moment.'

As we climbed the path by the side of the river, Faulkner explained to me how changes to farming practices helped bring about positive benefits. 'Fewer, better-quality stock result in a better price, less fertilisers in the ground save money and allow for more, better-quality habitat. If you undertake this sustainable approach to farming, not only does it solve a lot of the environmental problems, it creates a lot of environmental answers. We also saw how we should look for new products – environmental products – and that's where we started to tune in to this idea that managing floodwater is a product that farmers can produce.'

He showed me areas where works had been undertaken to arrest the disintegration of the blanket bog. At very low cost, blocks of degraded peat had been shifted into the streams that have appeared as the bog has fallen apart. In another area he explained to me how deciduous forest is being regenerated while in others the blockage of drains on farms alongside changed grazing practices were all beginning to deliver dividends.

The more the project has looked at what could be done, the more it became clear that multiple benefits could be achieved. 'We found we could address floodwater management, water quality, and we can even regenerate the bogs and start locking up carbon, creating climate change benefits. We realised you could change the whole thing around to create solutions to all of these major problems.' In pursuit of better water quality the removal of exotic conifers from the forestry plantations around the streams had brought about positive changes very quickly. The trees' needle-like leaves 'scavenge' airborne nitrogen pollution coming from transport, industry and farming and concentrate it. This gets into rivers, making them more acidic, and wipes out insect, fish and bird populations. The response during recent decades had been to build expensive plant that introduces lime into the rivers to counteract the acid. That requires power lines to run into the hills, costing millions of pounds.

An alternative approach trialled at Pumlumon is to remove the conifers along the stream sides. 'We take the conifers away and let natural regeneration take over. As a result we've got eco-logical corridors, which are fantastic for wildlife. The fish have come up here already'. He points to a little group of six-inch-long trout lying in a pool and tells me that salmon are back too. Then a dipper flies past. These handsome brown-and-white aquatic songbirds were lost from much of mid-Wales during the 1970s and 1980s because acidification of the rivers killed the water-dwelling insects they fed upon, but now they are coming back. 'So what was the cost?', I ask. 'A few chainsaw teams and a tiny impact on the forestry,' he says. From that investment not only is the wildlife and water quality improved, but the biggest economic benefit of all has been enhanced, namely recreation and tourism.

For all this to work and be economically sustainable, however, someone needs to pay for the new environmental products being produced, even if they're cheap. In this case some of the money needed for the upland restoration came from the charitable arm of Admiral Insurance. Faulkner commends 'an insurance company putting its money where its mouth is. As far as they're concerned they're interested in mitigating floods, and they could really see the bigger picture.'

He points out how a broader and sustainable solution would be to redirect the billions of pounds paid out to farmers each year through official subsidies. The potential for what could be achieved by doing that can be seen downstream in the Pont-bren catchment that drains into the upper Severn. Farmers here have taken it upon themselves to replant woods and hedgerows into the areas of improved pasture so as to provide the sheep with shelter and also a source of chipped wood that is a free source of animal bedding.

For the people of Tewkesbury perhaps the main thing that they'd be interested to know is how the simple and easy-to-repeat initiatives taken by these farmers have made a dramatic difference to how much water stays in the hills, and therefore out of their living rooms. If they knew the scale of difference being made, they'd very likely be pleased to see more of their taxes encouraging more of the same.

Field experiments have demonstrated how the soil beneath the trees holds much more water than in the pasture and the rate at which water entered the soil beneath the trees was found to be up to *60 times greater* than adjacent areas of grass. Researchers found that runoff reduced by up to 78 per cent on those sites planted with trees and hedges and concluded that this could be expected to dramatically reduce soil erosion. It was estimated that if other farmers in the catchment followed suit and planted 5 per cent of their land in this way then flood peaks downstream would be reduced by about 29 per cent. Significant beneficial impacts were measured within two to six years.

Other upland programmes are delivering similar benefits, including the blanket bog restoration mentioned earlier on Exmoor. Research carried out there by Professor Richard Brazier's team following the work to block up drainage ditches found a dramatic and rapid change. Using state-of-the-art equipment which relays data on, among other things, changes in water table from over 200 monitoring points on the moorland every 15 minutes, it was found that in areas where the bog is being restored *two thirds* less water is leaving the moorland during heavy rainfall compared with three years before.

When extrapolated across the whole 2,000 hectares of recovering moorland the results indicated that each year some 6,630 Olympic-size swimming pools less storm water would enter downstream rivers, thereby providing a significant buffer against flooding, including on the Exe, which in recent winters has contributed to chaos, misery and huge economic costs due to damaged infrastructure and flooding of homes of businesses.

The Upstream Thinking initiative that we heard about in the last chapter has also led to activities in the lowlands that will help reduce flood risk. Professor Brazier pointed out to me how the heavy clay soils in the lowlands had been damaged in ways that increased flood risk. 'Many of these soils are heavily compacted. You might see that as the root of the flooding, and water-quality problems, because there's just no storage capacity there. Unimproved grasslands hold on to water for longer and you can store up to five times as much as in improved grassland soils.'

Pete Burgess of the Devon Wildlife Trust explained to me how his organisation was helping farmers to turn this information into action by loaning out soil-aeration equipment across the Culm areas. The Trust is also advising on how best to improve farm infrastructure and putting guttering on farm buildings so as to help limit flash floods. Like so much of what can be done to manage flood risk it's inexpensive and not complicated, but does require leadership, partnership and good policies to gain scale.

When it comes to land use there is also considerable potential for using land explicitly for water storage at times of high flood risk. This approach has caused huge controversy in the wake of the floods on the Somerset Levels and elsewhere, but experience tells us that this is an important part of how we should be responding to the more extreme conditions that will accompany climate change. After all, a very common landscape feature used to provide widespread benefits in this regard.

Floodplain – a clue in the name?

Of all the different kinds of wetland that we have in the UK, from fen to blanket bog and from reed-bed to saltmarsh, the most extensive of all is floodplain, with nearly a million hectares of it spread around the country. These classic features of rivers in their middle and lower reaches, where the gradient toward the sea is gentler and the energy of the river is correspondingly lower, are an important part of natural drainage systems. When heavy rain falls in the catchment the river periodically spreads outside its channel and sediments eroded from the hills upstream settle as the energy that carried them thus far is dissipated.

The channel meanders across the bed of material brought down in centuries and millennia gone by, snaking back and forth, constantly reshaping the riverscape in an endless slow dance between sediment and water. The meanders that bend across floodplains increase the length of the river and slow down its flow. In more natural rivers woodland growing on the floodplain holds water back even more, as do snags such as fallen trees lodged in the channel. Or at least this is what natural floodplains do. In many river valleys the eternal dance has been brought to an abrupt stop.

With about 42 per cent of the floodplain in England and Wales physically separated from its river by engineered structures, including flood embankments, natural processes have been dramatically interrupted. The enormous effort required to tame rivers in this way has been for good economic reason. Many of our most important towns and cities are located on floodplains, not least because of the transport links provided by the rivers. Those sediments deposited from thousands of years of erosion are often very good for farming. As time has gone on, more and more floodplain has been occupied, especially during recent decades, when it seemed the needs of a growing economy were insatiable, and our ability to control rivers perceived to be without bounds.

This engineering and building was, however, undertaken without a full appreciation of the job being done by floodplains. A modelling study done on the River Cherwell in Oxfordshire demonstrates one of them. This river floods regularly and especially in winter. Research showed that separating the river from its various floodplains by embankments would increase peak flows downstream by 150 per cent. And when parts of downstream are heavily urbanised, the consequences of removing this water storage function can be painful.

Another consequence of separating rivers from their floodplains is to change the energy in the water flow. Instead of flooding across flat areas the water runs through quickly, taking its sediment load with it rather than leaving it behind under the relatively calm waters of its floodplain. Assuming the river does eventually slow down, then the sediment it is carrying will drop out, making the riverbed higher, thereby increasing flood risk somewhere else. Thus, many of our floodplain defences simply move flood risks downstream.

Then there is the question of where the sediment is coming from in the first place and how much of what we do with land makes clogging up of watercourses worse. As we have seen, many of our soils have suffered different kinds of damage, and not only has this contributed to lower water quality and led to nutrient build-up in the environment, it can also exacerbate flood risk. When soil is no longer on the land and instead suspended in river water it has several possible fates. It can end up in the sea, a water treatment works, be deposited on a floodplain or be left on the bed of a slower-flowing stretch of the river. None of these are desirable, but all are increasingly the case because of the way we farm. When coupled with the more extreme conditions that will accompany climate change, then even worse consequences can be expected.

In January 2014 I was shocked to see pictures of deep gullying of soil on a mid-Devon hillside near the village of Sampford Peverell. Because of heavy rain falling on bare soil the water began to flow in little streams. These became bigger streams, and then torrents. Soil that took thousands of years to build up was torn away in a single night. Some of the gullies were nearly two metres deep and looked like miniature Grand Canyons. I have seen that kind of thing often enough in highly degraded tropical regions, but never in the southern English lowlands. Once the soil had left the field it was of course on its way to the rivers.

Machinery and animal hooves have also caused many soils to become compacted, so that instead of absorbing water, holding it and slowly letting it go, it runs off hard impermeable surfaces all at once. And, as we also saw in Chapter 1, because of intensive farming there has also been a widespread reduction in the proportion of soils comprised of decaying plant remains – so-called organic matter. Soils depleted in this way hold less water. The next time you see a river in high flow or flood after heavy rain, note what colour the water is. Across much of the UK it will be that of cocoa – or more accurately soil.

Perhaps this is the point to pause and consider the approach of Iain Tolhurst, the organic farmer we met in Oxfordshire in Chapter 1. He makes every effort to ensure there is no bare ground on the land he farms, not only to conserve nutrients, but also to slow down erosion. It seems he might have a message that those who determine how farm subsidies are spent could benefit from hearing.

When it comes to clogged rivers the effects of farmland erosion can be compounded by over-abstraction of groundwater. Charlie Bell, who we met in the last chapter working to restore the River Beane in Hertfordshire, has seen how lower flow paradoxically contributes to the river being more flood-prone. 'The river becomes much more silted and gets clogged up with vegetation. I've had conversations with farmers whose land is flooding as a result, so the service the river once provided in taking water off the land has been reduced because it has become silted up. A functioning river with the right amount of water helps to reduce flooding.'

The growing realisation that more natural floodplains actually have a lot of value has led to actions for them to be restored. Some of this is about 'rewetting' land so that flooding is allowed to happen. Although often controversial and against the 'drain-and-dredge' culture that we have become accustomed to over the decades, it can work well – and at relatively low cost.

Nature's storage tanks

A mosaic of reed-fringed water, marsh, wet woodland and scrub known as Potteric Carr occupies the southern edge of Doncaster. It covers around 200 hectares and is looked after as a nature reserve by the Yorkshire Wildlife Trust. The result of subsidence caused by coal-mining activities, the shallow bowl that holds the wetland habitats here is a magnet for wildlife. Among its more notable features are breeding black-necked grebes and avocets. While its primary purposes is conserving wildlife and offering a place for people to relax and enjoy the outdoors, there is another reason why the residents of the adjacent South Yorkshire city should be pleased to have it there – protection from floods.

In November 2007 heavy rainfall on the Pennines caused a vast quantity of water to enter the River Don. A great pulse ran downstream, causing extensive flooding in Sheffield before speeding on to inundate much of Doncaster. Sitting at the very northern fringe of the flat East Anglian landscape, the city is in another of those low-lying parts of England that have for centuries been subject to drainage works so as to permit farming. Cornelius Verymuyden, the Dutch engineer who designed the famous system of drains and washes in Cambridgeshire, was also behind the drainage works here. A linchpin in his plan was a feature known as the Mother Drain. It was a classic Dutch system, whereby fields were emptied by gravity of excess water, which flowed into ditches before being lifted by wind-powered pumps into rivers that in turn took the water to the sea.

As we saw earlier, the drainage of peaty soils causes them to shrink so that land once at sea level or just above contracts over time to eventually lie beneath it. This has happened over extensive parts of eastern England, including around Doncaster, and has resulted in rivers that are *above* the surrounding land. Banked up over centuries, they take water pumped up from the adjacent land that gets ever lower as the peat disappears.

The Mother Drain comes out of south Doncaster and water collected in it is pumped into the River Torne. Water from the Torne is then pumped into the Trent, which in turn drains into the North Sea via the Humber Estuary. It generally works pretty well and flooding is rare. But when the system gets overwhelmed there can be problems.

Rob Stoneman, Director of the Yorkshire Wildlife Trust, explained what happened in 2007: 'It's effectively a bath tub without a plughole, as we saw with the catastrophic event. We had intensive rains over the Pennines, probably exacerbated by peat-bog drainage, the River Don burst its banks and the whole of the centre of Sheffield was under water, including the Meadowhall Shopping Centre. Then the water went out along the lower Don towards Doncaster, again pouring over into the low-lying ground which then filled up to a metre or so deep.'

As the capacity of the ancient drainage system was exceeded by the sheer volume of water, a massive human and financial cost was incurred. Rather than draining out in a day, which is what would normally happen, it took weeks to clear. Some families spent two Christmases living in caravans while about one billion pounds was paid out in insured losses.

But while north Doncaster was heavily affected by flood-water, the south side of the city wasn't. One reason was because when the Yorkshire Wildlife Trust designed their Potteric Carr reserve they did so with floodwater storage in mind. Stoneman again: 'The reason why south Doncaster didn't flood was partly because of the catchment, but largely to do with the fact that it has massive water-storage facility. Potteric Carr has a whole system of drains and dykes, and if any area floods it can store a heck of a lot of surface water. The visitor centre and a few hides get wet, but it's all pretty resilient, so it doesn't actually matter that it fills up with water. We had to shut the site for two weeks and people couldn't go in there, but it gradually subsided. How much did that save south Doncaster? That was never assessed. But if you put into context the cost of the south Yorkshire flood at about £1 billion, then not having those people's shops, houses, offices flooding, must have represented millions, if not tens or hundreds of millions.'

Further examples of flood-risk management being combined with conservation and recreational values can be seen across Britain. For example in early 2014 people living near the stretches the River Itchen that are being restored, as discussed in the last chapter, saw how ecological improvements there have not only brought aesthetic, wildlife and sporting benefits. In February the natural areas helped protect homes and businesses from flooding. The sustained heavy rain that fell then would have caused the inundation of homes and businesses, were it not for water being held back in the river and on its recovering water meadows.

Then there is the case of the Insh Marshes, a wetland nature reserve that provides flood benefits to the Scottish resort of Aviemore. The marshes cover about 1,100 hectares of floodplain and have a role in water storage that has been valued at £83,000 per year (assuming the alternative was to replace it with seven kilometres of flood defence banks). The Marshes contribute to the local economy in other ways as well, and for example are worth an estimated £132,000 per year in tourism, whilst fishing adds a further £35,000.

These and many other cases demonstrate how wetland conservation, restoration and creation, the renaturalising of rivers and planting trees and hedges in the right places, can all help to reduce flood risk, and often in highly cost-effective ways compared with dredging and hard defences alone. A once widespread species of native mammal might help too.

Engineers in fur coats

The Knapdale Forest in Argyll, western Scotland, extends across a large area of mixed woodlands, lakes, grasslands, hills, sheltered sea lochs and coast. In the north of Knapdale and about three miles south of the coastal hamlet of Crinan is a long narrow freshwater lake called Loch Coille-Bharr. It lies in a striking landscape. There are extensive areas of coastal rainforest dominated by a mix of deciduous trees including birch, field maple, oak, hazel, rowan and ash. There is a rich understorey of shrubs, a variety of wild flowers and many of the trees are festooned with green and grey coverings of lichens and mosses. Set in this moist forested landscape on the eastern side of Loch Coille-Bharr is an unusual wetland, a type of natural feature not seen in these islands for centuries. It recently appeared because a stream that once ran into the small Lock Dubh and out again into the bigger water body became blocked.

Where the stream once made its exit there is now a barrier made from cut branches and rocks. Mud has been plastered into the cracks. On top rushes grow, their roots helping to bind the structure together. The obstruction is about ten metres long and has considerably increased the size of Loch Dubh, in the process causing an extensive shallow wetland to appear.

It's the first week of May and a newly arrived wood warbler delivers its stuttering song from a copse on the far side of the shallow water. In the warm shallows a multitude of little dimples are created by schools of baby fish. Insects buzz around the drowned tree stems. A frog croaks. As it gets dark groups of bats begin feeding around the margins of the big pool, catching winged insects in the still temperate air. A tawny owl hoots from a distant copse and as the evening gloom gathers I strain my eyes to see if I can catch a glimpse of those responsible for making this place look the way it does.

Short of a direct sighting there are plenty of clues as to what is going on. Among the heather and blueberry thickets that separate the two lochs there is a network of neat trails. A line of recently flooded trees have been cut at the base of their stems and fallen in the same direction. A patch of birch saplings have been felled and cut branches lie scattered around. The base of the trunk of a very large mature tree has been expertly sculpted to the point where the removal of a little more wood will cause it to fall over into the water. Big wood chips like those that would be produced by an axe lie around the graceful curves of the recently exposed core of the tree. This is not the work of a lumberjack, however. No doubt about it: beavers.

The industrious animals have engineered the environment here so as to improve their food supply, security and housing. The dam blocking the outflow of Loch Dubh is integral to the whole plan. By raising the water level the beavers can swim between their food sources of twigs and bark and also float branches around more easily than carrying them. The water level has been elevated to cover their submerged lodge entrance and increases the number of trees in or near the water. So as to extend their foraging zone they have also dug canals.

These remarkable semi-aquatic creatures are rodents, so belong to the same group of mammals as mice, squirrels and guinea pigs. They are rather bigger, however. European beavers weigh in on average at about 18 kilos (that's about the same size as my springer spaniels), with the largest reaching 32 (about the same as an adult Labrador). These big animals, with their dense warm brown fur and characteristic stiff flat tails that propel them through water, once lived in wetlands throughout Britain. Hunted for their meat, pelts and scent glands, the last beavers in England and Wales were killed during the twelfth century, with some holding on in Scotland until the sixteenth.

Then in May 2009 four pairs of beavers were released into that forested lake landscape in Argyll. They were caught by expert trappers in the Telemark region of Norway where the terrain is similar to that of the British release site. The Norwegian beavers were believed to be the closest relatives of those that once lived in northern Britain and thus identified as the best source animals for reintroduction to Scotland.

The official Scottish Beaver Trial was initiated by The Royal Zoological Society Scotland and the Scottish Wildlife Trust and ran for five years through to May 2014. The results will be reviewed by the Scottish government so as to determine whether further releases should go ahead. While at the time of writing that process is not yet complete, from what I saw in that wooded wetland in Argyll during the final month of the trial, I'd say a profound positive impact has been created.

There are several reasons why the return of the beaver might be a good idea. In addition to improving the conditions for other wildlife, people love beavers and seeing them in a natural setting creates enormous pleasure. They attract visitors who bring tourist revenue. Then there is the impact beavers create through engineering their environment – including their dam-building activities.

In the wake of the 2014 floods the Mammal Society, an organisation devoted to the conservation of our native mammals, called on the government to consider the role of beavers in its overall future flood-defence strategy. Harnessing the animals' remarkable engineering skills to renaturalise rivers, the idea would be to use these creatures in flood-prone river systems to hold more water in the environment for longer.

It could be part of a plan that would look again at the efficacy of the approach used by engineers in hardhats, straightening and clearing rivers so as to get rainfall off the land and into the sea as swiftly as possible – the downsides of which include the massive expense of dredging and other maintenance works, not to mention the fact that it sometimes doesn't work, as we found out in early 2014. Those human 'solutions' also contribute to that brown fringe of soil we can see in the sea around our coasts from space. Through the construction of dams and canals the beavers change the dynamics of rivers. They reintroduce steps along river profiles, slowing down the discharge of water and catching sediments. The dam ponds also catch nutrients and carbon and incorporate these into plant-rich 'beaver meadows', in the process not only helping to increase wildlife diversity but reducing the need for expensive water-treatment equipment downstream.

The habitat diversity created by beavers increases wildlife richness, including among invertebrates, birds, mammals, reptiles, amphibians and fish, with the latter leading to new angling opportunities. No wonder some ecologists refer to these animals as 'ecosystem engineers', and no wonder that since the 1920s some 24 European countries have reintroduced beavers.

There are now quite advanced proposals for the release of beavers into the wild in Wales and an in-depth study co-authored by the government's official conservation advisors concluded it would be positive for them to be returned to suitable locations in England. In advance of such releases, though, there have already been deliberate attempts to copy beaver-modified habitats so as to reduce flood risk. The Environment Agency has for example worked with locals around the village of Belford in Northumberland to mimic the wetlands, storage ponds and dams that are created by beavers. The idea was to slow the flow of water downhill after rains, releasing it more slowly and thereby reducing the size of flood peaks. It seems to be working.

The many attempts to restore wetlands, improve soils and to create storage areas for floodwater show how such a mindset is already delivering benefits in a cost effective manner. Considering what we know about the likely consequences of climate change, and the ever more extreme events that are coming with it, it seems that learning from these examples and taking them to greater scale is a fast-emerging priority.

So what of the politicians and commentators who urge that we 'dredge, dredge, dredge'? While this message has the merit of being simple, there are two questions that must be answered.

Namely, will it work, and might there be better things we could do? When it comes to the first of these, the answer is 'it depends'. But even in places where some relief might be found by making rivers straighter and deeper, the benefit is likely to be only temporary. After all, if the source of the problem has not been dealt with then it will only come back again.

On the second, the answer is an emphatic 'yes'. This is not only because dealing with the source of the problem is in the end a more durable option than treating the symptoms, but also because it could bring so many other benefits. These include cleaner water, carbon storage, food security, energy saving, the conservation and restoration of wildlife and enhanced health and quality of life.

An equally broad range of benefits could come through working with nature to reduce the risk of coastal flooding.

The angry sea

The UK has a long coast, one that in many places is vulnerable to inundation from the sea. On the night of 31 January 1953 the people of eastern Britain found out just what that can mean.

The disaster began to unfold when a deep low-pressure system formed over northern Scotland, not only whipping up gales, but also sucking up the sea surface, to an extent that the tide didn't go out that evening. At the same time, the violent winds forced billions of cubic metres of water south into the funnel-shaped North Sea, further raising the sea level. With a vast amount of water already backed up, the tide turned toward the highest spring tide then known, in some places rising 5.6 metres above the average.

Water spilled between coastal dunes and overtopped sea walls right along the eastern coasts of Scotland and England.

People had gone to bed without warning as to the impending risk and many were awoken to see under the full moon a white 'landscape' that at first they believed was snow, but which they soon realised was waves crashing across fields and into their streets. Along the southern North Sea coast from Lincolnshire to Essex and into the East End of London, floodwaters in places ran up to two miles inland from the shore. For some the first they knew of the danger they were in was when the door crashed open and the sea came inside their home.

More than 1,800 Dutch citizens were killed in the 1953 floods, while in England 307 people lost their lives, and a further 19 in Scotland. One of the worst-hit places was Canvey Island in Essex, where 58 people perished and 21,000 were made homeless. More than 1,600 kilometres of coastline were damaged and over a thousand square kilometres of land were flooded.

In the wake of this disaster, the UK and Netherlands planned extensive coastal defences including many barriers and dams. These have repeatedly proved their worth, for example at the end of 2013 when a surge even bigger than that in 1953 caused only minor damage. Whether our existing defences are enough to withstand future risks is, however, less certain. A 2014 Met Office report that made a link between the extreme weather early that year and climate change also issued a warning as to the growing peril posed to our coastal areas. It pointed out how sea level has already risen during the twentieth century due to ocean warming and melting of glaciers and concluded: 'With the warming we are already committed to over the next few decades, a further overall 11-16 cm of sea level rise is likely by 2030, relative to 1990.'

Managing the risks that will come with this will certainly require engineering interventions. At the same time though vital roles must be played by natural coastal features.

Natural barriers

Many different kinds of ecosystems help stabilise Britain's coast and create natural sea defences, including reefs, seagrass beds, mudflats and saltmarshes. All these form natural barriers, while sand dunes also act like gigantic sponges soaking up excess water. Many of these natural systems also self-adjust as the sea-level rises, making an even more cost-effective contribution in how we might cope with change. To make the most of what nature can do in protecting coasts will require a reversal in historic trends, however, and for the progressive loss of natural shorelines to be replaced by programmes that lead to their regeneration and sometimes re-creation.

Indicative of the growing realisation of the role that might be played by nature in defending our coast is how defences erected after the 1953 floods are now being removed so that the coast might return to a more natural state. One such place is along Freiston Shore on the Lincolnshire coast, where old sea defences have been taken down and new ones, in the form of coastal wetlands, installed instead.

John Badley is the site Manager at RSPB's Lincolnshire Wash Reserves. He told me how the creation of the managed realignment is part of a plan 'to work with nature and not against it'. He explained how with increasingly volatile conditions linked to climate change it will be 'very difficult to hold the line' with engineered defences. In facing the imminent risk of rising sea levels, hard defences such as sea walls will become both more impractical and more expensive to maintain.

Recognising the growing challenge, in August 2002 the sea wall was breached in three places at Freiston Shore and the sea allowed to flood onto the landward side. Within a few years most of the newly flooded area, covering about 66 hectares, had returned to saltmarsh. The new marshes have already brought a number of benefits. The threat of flooding has been reduced to homes in and around the Lincolnshire town of Boston, while some 80,000 hectares of low-lying Grade 1 agricultural land are more secure. This benefit arises from how saltmarshes reduce the power in waves. One study found how over a distance of 40 metres saltmarsh reduced the height of large waves in deep water by nearly a fifth, making them an effective tool for reducing the risk of coastal erosion and flooding. Sixty per cent of this reduction was found to be due to the presence of the saltmarsh plants.

Prior to the construction of the new defences the engineered bank provided protection against the kind of event that would be expected to occur about once in each twenty year period (over the course of a year therefore a 5 per cent risk of breaching). The new scheme is estimated to provide protection against one-in-two-hundred-year events (that is a 0.5 per cent risk of breaching). And the enhanced protection has been achieved at modest cost. By dissipating wave energy (rather than trying to hold it back with a wall) the new wetlands have reduced the need for capital-intensive defences and are estimated to save about £3,000 per kilometre per year in maintenance costs compared with hard defences.

Then there are the co-benefits. The new wetlands are important nursery areas for fish and shellfish, and they support large numbers of wintering geese, duck and waders. That in turn has proved a tourist attraction, attracting over 43,000 visits in a year. Badley told me how the RSPB's reserves here are now 'in the top twenty Trip Advisor best places to visit in Lincolnshire, up there with Lincoln Cathedral'.

Wilder world

Climate change used to be discussed as though it was some kind of theoretical future threat, a matter of faith or an abstract idea to be debated separately from reality. That is changing. For while single events can rarely be attributed to climate change, the ever-increasing number of them, one after the other, increasingly can be. Repeated record-breaking extremes, from floods in Pakistan to heatwaves in Russia and from extreme cold in the USA to drought and fire in Australia, provide graphic warning that something major is under way.

Here in the UK we have recently seen our own rash of out-of-the-ordinary weather. And it's not only about flooding and the consequences of too much water. The 2003 heatwave that hit Europe caused about 30,000 premature deaths. In 2011 parts of the UK were in the grip of a record-breaking drought. That year the Centre for Hydrology and Ecology reported the driest spring for a century, leaving fields parched and rivers with record low flows.

Farmers were hit worst, but it was feared other sectors such as food processors, energy companies and some manufacturers that relied heavily on taking water from rivers or underground sources were expected to experience problems as the amount of water they would be allowed to take would need to be cut, thereby affecting productivity.

By the autumn of 2011 in some parts of Eastern England river flows were among the lowest ever recorded, lower even than during the extended drought of 1976. The exceptionally dry conditions continued into 2012 with water companies expressing alarm at the prospects for summer water supply. Leading figures in the industry said that only exceptional rain could avert a crisis. We were lucky, because that is exactly what we got. Starting in April 2012 it began to rain, and it rained to the point where the year ended as the wettest ever for England and the second wettest for the UK, just 6.6 millimetres short of the previous record set in 2000. Many areas were affected by the extreme rainfall, with thousands of homes flooded and farmers struggling to grow crops in the saturated soil.

That wet summer appears to be part of a trend in which extreme rainfall is increasing. One indicator is Met Office data showing how four of the top five wettest years ever recorded in the UK were from 2000. Then came the deluges of the winter of 2013-14, bringing one of the most exceptional periods of rainfall in England and Wales since records began 248 years before. Unsurprisingly there was a renewed debate as to whether extreme weather conditions could be linked to climate change. The answer from the Met Office's Chief Scientist Julia Slingo was very clear: 'all the available evidence suggests there is a link to climate change', she said at the launch of a report on the extreme weather released in February 2014.

Previous to this it had become fashionable with climate sceptics and many media commentators to repeat the mantra that no one event could be put down to climate change. As is the case at the global level, it was not so much the individual events that caused the Met Office to make the link, but the fact that it was so unusual to see so many storms one after the other in a short space of time.

The report also said there was evidence the exceptional winter could be part of a longer trend, with the number of winter storms in the North Atlantic between the UK and Iceland on the increase, and on average becoming stronger. All this fits in with the scientific understanding of the fundamental physics of a warming world.

Computer-based tools that can attribute the likelihood of any particular event being caused by human-induced climate change are yet to confirm exactly how much of what we're seeing is down to human influence on the climate, but that will soon be available. In the meantime it would be the height of folly for any country to continue as if the jury was still out – and not only in relation to flooding, but also storms, heatwaves and droughts.

When the Prime Minister stood up in the House of Commons in January 2014 and suggested that the unprecedented conditions facing the country were linked to climate change his remarks were greeted with groans of derision by many of his own MPs. Considering the overwhelming evidence as to the growing peril at hand, this wilful disengagement is a serious abdication of responsibility. After all, if our politicians are unable to ensure the security of the country and its citizens, what are they there for?

Many of them would say that economic growth is their priority, and that looking after the interests of business is top of their list. If this really were the case, however, then surely they should be paying attention to the message coming from the insurance industry? London is after all the world's leading insurance centre and the sector is vital to the future health of the City. In common with the meteorologists, insurance companies rely on expert number-crunchers who base their work on facts and measurements; the success of their businesses relies on the accurate calculation of risk, and these companies know that risk is changing, because the climate is.

As payouts linked to extreme weather events have soared, so insurers have deemed it unwise to insure properties evidently at high risk, such as those located in floodplains. In the face of a changing constellation of risk the insurance industry really has three options: to withdraw cover, increase premiums or ask taxpayers to take on risk via government guarantees. For households and businesses none of these are particularly desirable. The fact that so many political leaders and representatives remain in denial as to the ultimate causes for the changing nature of risk only makes it worse. Were they to accept the facts, they'd be required to not only adopt a programme to reduce emissions of climate-changing gases (we'll come to that in the next two chapters) but also one that prepared the country for what we can expect to be a more volatile future – a future where what previously seemed like the exception becomes the norm, and where the unprecedented enters our collective experience.

Considering the changes we can see coming down the track, the adoption of a coherent and thought-through plan to prepare the country to be resilient in the face of extreme conditions is obviously required. It is fair to say that we don't have that yet, and less still the means to implement it even if it did exist. Deep cuts to the budget of the Environment Agency, which among other things has responsibility for flood protection in England, have been presented as a means to remedy the country's huge financial deficit. At the same time official figures showed how during the five years up to 2013 the go-ahead was given to build some 4,000 homes on floodplains against the advice of the Environment Agency. In another attempt to cut 'red tape', save money and promote economic growth, Communities Secretary Eric Pickles in 2010 scrapped the obligation on councils to prepare scenarios for how climate change could affect their area.

Planning for resilience

Despite the recent tendency for policy to be in reverse gear, there is fortunately evidence for more rational responses emerging in different parts of the UK. Water companies, farmers, conservation groups and partnerships between official, voluntary and business organisations are all delivering practical programmes that will help soften some of the blows that will inevitably come with further climate change. It is being done at modest cost, too. For example South West Water's programme to help restore some 2,000 hectares of degraded blanket bog on Exmoor will, between 2010 and 2015, cost the company (and therefore its customers) just £2.2 million. As we saw, this happens to be a co-benefit, coming with a programme to cut costs in cleaning up the water ready for public supply.

It's but one of a growing number of initiatives that demonstrate how it is generally better to focus more on flood-risk management compared with static defences. Dykes, dams, barriers, drains and dredging will of course all be needed in different situations, but they will be more effective if used in a more integrated strategy that also takes account of what nature can do if we maintain or restore its functions.

This is not necessarily an easier thing to do, however. Designating areas of farmland for floodwater storage, rather than trying to keep it dry, is a controversial idea. So is the creation of wetlands and river restoration in place of dredging. There is now, however, a great deal of experience from this country and others to demonstrate the economic sense of going down this route. Tools include altering the way farm payments are made, so that those who look after the land are encouraged to reverse soil compaction and restore bogs. There could be new incentives to encourage tree and hedge planting, and compensating farmers for their land being used for flood storage. These and other measures could be put in place with a bit of imagination and joined-up policy-making. New guidance for the water companies would help too, as would a more integrated use of the money being spent on reducing flood risk.

More and more people can see the sense of an integrated approach. One is Somerset Levels resident Stephen Moss. He's a lifelong naturalist and was the original producer of the BBC's highly successful *Spring Watch* series. After he took me to see the flooding on the Levels in early 2014 he told me how he'd come to appreciate the benefits of a joined-up approach. 'The flooding taught us an important lesson, one that our ancestors knew very well, and that is how we need to work with the land

and water, not against it. This means a more sensible approach toward farming, with less intensive land use to allow water to filter slowly down towards the sea, an acceptance that expensive engineering solutions are only a short-term measure and could even make the problems worse in the longer term. We have an opportunity to create a truly special place for people and wildlife, a place where visitors will come from far and wide to enjoy the wonders of this beautiful part of Britain'.

There is another major benefit that can come through working with the grain of nature – it is linked with how best to keep climate-changing gases out of the air in the first place. It's time to head north, to the wild mountains of Scotland, to find out how.

Manifesto

- **Require public bodies** to collaborate in drawing up green infrastructure strategies that identify those areas where tree and hedge planting, wetland restoration, natural coastal enhancement and renaturalising of floodplains would bring most benefit in reducing flood risk and target funding accordingly.

- **Require water companies** to undertake economic studies to identify where it would be cost-effective for them to foster partnerships in catchments so as to improve water quality and cut flood risk while reducing bills. Specify this in the context of a new water-stewardship duty on water companies to sustain the whole water cycle – not just in terms of water supply and sewage treatment.

- **Produce maps** to indicate where it makes sense to periodically store floodwater on the land and provide official advice and assistance to farmers in integrating that function with food production. The Somerset Levels would be a good place to start.

Nowhere to run to – Scotland's mountain hares are among the British species most at risk from climate change.

Chapter 6
Carbon country

THREE TO FIVE MILLION YEARS AGO – THE LAST TIME CARBON DIOXIDE CONCENTRATIONS WERE AS HIGH AS THEY ARE TODAY

80 PER CENT – REDUCTION IN GREENHOUSE GAS EMISSIONS REQUIRED BY 2050 COMPARED WITH 1990 LEVEL

ABOUT DOUBLE – CARBON CAPTURED IN PEATLANDS COMPARED WITH THAT IN FORESTS

It's mid–May and I'm heading up a track on the side of a steep-sided glacier-carved valley at the head of Glen Prosen in the Cairngorm Mountains of eastern Scotland. As I climb higher the path crosses from grassy slopes to those covered with heather. The land becomes wilder, and wetter, and the path indistinct as I find myself sloshing through peaty water.

When the steep climb levels out, I look across to the north and see that I'm walking on a wide undulating plateau. The map says that I am at about 700 metres above sea level and in the distance high rounded peaks are set against a deep blue sky. Nearby are patches of snow scattered in north-facing hollows. Further away and a little higher up are unbroken snowfields. A movement to the right catches my attention and my binoculars alight on a mountain hare. The exquisite creature bounds ahead, stops, turns and looks back at me, ears bolt erect. This Ice Age refugee is an animal of tundra-like habitats and Cairngorm is a British stronghold.

In transition from its pure white winter coat to grey-brown summer fur, the hare is mottled with grey and pale patches. It is a perfect match with the early summer landscape, where winter white mingles with the greys and browns of lichen, heather and moss. The still pure white fur on its feet and legs look like new long socks.

As I wander further out across the high open moorland, there are more mountain hares. About ten of them gather to have a look at me. For once the wildlife seems as interested in me as I am in it. My coterie of hares is joined by a ptarmigan – a kind of Arctic grouse. The bird flutters onto a grey lichen-encrusted rock and makes a gobbling sound as if to warn me off his land. Another Ice Age relict it is, like the hares, in transition to summer colours. In winter these birds are the colour of snow and at this time of the year its densely feathered feet and underneath are still pure white, but on top it now has different shades of grey. A bright red eyebrow denotes a male bird.

A tiny patch of snow next to the rock on which the bird has landed confirms the utility of its camouflage, and how its plumage is not only beautiful, but also brings essential survival value. If ever there was a suggestion that aesthetics and practicality need to be in conflict, this bird dismisses it.

Then a red grouse flushes from a hollow, making a noisy fuss as it flaps and glides a short distance to land on a little ridge. It settles in a patch of heather atop of a low black cliff about three feet high.

As I focus on the stout bird I notice that the vertical black face of bare ground below where it sits is made of peat. Looking around I suddenly realise there are quite a lot of bare black patches, where the heather and the otherwise dense wet carpet of *Sphagnum* moss is sparse or non-existent. The bird is on a little moss- and heather-topped island in a sea of black peat. Then I see that the moorland is covered with little patches of elevated peat, denoting the former land surface and revealing how the ground in between had disappeared.

It was not obvious why the peat was disintegrating, but most likely it was down to a combination of pressure from grazing animals (sheep) and fire. The ecosystem I found high up there in the Cairngorms was broadly the same as I saw on Dartmoor and in the middle of Wales at Pumlumon. Although widespread in the UK, this blanket bog is rare in a global sense. About a fifth of Scotland is clothed in it, which is about 15 per cent of the global total. Some areas of blanket bog in Caithness and Sutherland, the so-called Flow Country, are among the most extensive and intact examples anywhere in the world.

The cool blanket

Blanket bog is by far the most extensive of the UK's habitats. There is about 1.5 million hectares of it, mostly in Scotland but with about 355,000 hectares in England. Most of it began to form around 5,000 to 6,000 years ago, due to a combination of the clearance of the original forest by Mesolithic and Neolithic people and a change toward a cooler and wetter climate.

These seemingly wild and untouched areas are, however, showing far more widespread signs of stress than that visible during my hike into the hare and ptarmigan habitat in Cairngorm and that I saw in mid-Wales at Pumlumon. One recent survey suggests that about a third of Scotland's total blanket bog is eroding while in Wales three-fifths of the blanket bog is described as 'modified'. The situation in England its even worse, with only about 4 per cent of upland peatland deemed to be in a favourable ecological condition, where mosses are still actively forming peat. Around half of the total area has been modified to the extent that little or no peat-forming vegetation remains, including because of burning to encourage grouse. As we have already seen this can present problems for water quality and flood risk. It is also a major source of carbon dioxide going into the atmosphere.

As I looked at the ground around me on that sunny day on those stunning Scottish hills it seemed the peat was not eroding only by being washed downhill after heavy rain. It was certainly leaving via that route. But it was also evaporating. With the wet mossy surface broken up, water had run out of the surface of the peat, leaving it drier and exposed to the air. When this happens the carbon captured in dead plants thousands of years ago combines with oxygen in the atmosphere to make carbon dioxide. As the organic material oxidises, so the land literally turns into thin air.

This release of carbon from the ground and into the atmosphere is part of a trend that, as we saw earlier, is also under way in agricultural soils, where cultivation is reducing the amount of organic material. The worldwide degradation of ecosystems like that Cairngorm blanket bog is one reason why on 9 May 2013 for the first time in at least 800,000 (and probably more like three million) years the concentration of carbon dioxide in the Earth's atmosphere reached 400 parts per million.

The last time carbon dioxide concentrations reached this level was most likely during the Pliocene epoch, between three and five million years ago. Recent estimates suggest that CO_2 levels back then reached as high as 415 parts per million – a level that on present trends the Earth will once again experience during the early 2020s. In the Pliocene, that concentration of carbon dioxide was accompanied by global average temperatures 3 or 4 degrees centigrade warmer than today's. The Arctic was nearly ice-free, trees grew in Greenland and sea levels were up to 40 metres (more than 130 feet) higher than today. As we load still more greenhouse gases into the atmosphere the scale of change that we can expect to follow is profound, and the higher we push it the bigger the difference we'll likely experience.

This is a problem at a number of levels, including for Britain's mountain hares and ptarmigans. In common with other animals and plants confined to the small areas of Arctic-like wilderness in our islands, the gravest threat they face comes from climate change. Some wildlife species can move north or uphill in response to warmer conditions, but when you are already at the top of the hill in the north of the country, there is nowhere left to go. This is why their fate is in part dependent on the carbon that is held in the habitats they live in.

Due to the release of greenhouse gases caused by human activities the global average temperature has already risen by about 0.8 degrees centigrade compared with the time just before the industrial revolution, and is projected to continue rising as more and more heat-trapping gases are released into the atmosphere. If we carry on as now, some projections suggest that by the 2060s the average temperature could increase by 4 degrees compared with the time before industrialisation.

With progressive warming come rising sea levels. This is presently running at an average of about three milimetres per year, with one third of that caused by the melting of the Greenland and West Antarctic ice sheets. The rest comes from mountain glaciers and the expansion of the oceans as they warm up. There has already been a 20-centimetre rise since pre-industrial times and it could go about a metre more by the end of the twenty-first century. If either of those two massive ice sheets becomes unstable and undergoes more rapid disintegration, which they could do, then it might be a lot more than that. In any event the rate of sea-level rise is speeding up.

A warmer world also has more energy in the atmosphere and oceans, thereby kicking off and sustaining more powerful storms. And with each one-degree increase in average temperature there is about 7 per cent more water in the air. More water in the air means more falling as rain. The more extreme conditions will also lead to more droughts as well as flooding. All this will cause ever more costly impacts to human societies, in both humanitarian and economic terms.

These trends are sometimes hard to grasp in a day-to-day sense, especially when specific events appear to contradict them. For example, a cold snap in London, where most of the UK's national media and politicians are located, can easily divert our attention from the gradual overall warming. We had one of those in early 2010. But as some sceptics seized on this as 'evidence' that the world was not in fact warming, it was a very local state of affairs that conveyed this mistaken impression. While London was a degree or so below freezing, in southern Greenland it was a balmy 10 degrees centigrade *above* freezing – an almost unheard-of high winter temperature for that Arctic latitude.

And it's not just the atmosphere that's becoming warmer. A substantial proportion of the heat trapped by greenhouse gases we've released has been absorbed by the oceans.

As this worrying warming trend has become more apparent, so an intense debate has broken out on how best to reduce the emissions of the greenhouse gases that are causing the climate to change. This is not a small thing, requiring cuts in carbon dioxide and other warming agents amounting to an 80 per cent reduction by around 2050, compared with 1990. This will need to be done at the same time as the world population increases by two billion, and while countries worldwide continue to look for new ways to grow their economies.

When it comes to the question of how to make these kind of reductions, the main focus has been on energy. That's clearly vital. However, far too little attention has gone toward reducing greenhouse gases coming from changes taking place in the natural world.

For the UK, this is significant, as our landscapes have a great deal of carbon tied up in them, especially the country's extensive peatlands. Across the UK as a whole it's estimated that between about 3.2 and 5.1 billion tonnes is stored in peatlands, making this the single most important ecosystem carbon store in the country. Most of it is in Scotland and Northern Ireland, with an estimated 475 million tonnes locked into English and Welsh fens and bogs. (If released into the atmosphere, each tonne of carbon, when it unites with oxygen, makes more than three and half tonnes of carbon dioxide).

To put this into perspective it helps to visualise what that carbon would look like as a pure gas in the atmosphere. One way to do this is to imagine one tonne of pure carbon dioxide as a spherical bubble about ten metres across (twice as high as a double-decker bus). Billions of bubbles that size obviously make a difference to the atmosphere, and that's why its best the carbon stays locked up in ecosystems, such as in that hare and ptarmigan habitat.

Peatland only covers about 3 per cent of the Earth's land area but it is estimated to contain 30 per cent of global soil carbon (around 550 billion tonnes), which is about double the carbon stored in all the world's forests. So although the UK might see itself as a small player in the global land-carbon debate, for example because it is a small country and doesn't have a high current deforestation rate, it does have a lot of peat. That means a big role in shaping carbon emissions through how we use the land. According to one estimate, damaged British peatlands are releasing almost 3.7 million tonnes of carbon dioxide each year, equivalent to the emissions of the combined households of Edinburgh, Cardiff and Leeds. And it is probably an underestimate as it excludes certain kinds of peatlands such as the East Anglian fens. A more comprehensive analysis from Wetlands International suggests it might be nearly three times this amount.

There are many reasons why Britain's peatlands are being degraded. Large areas have been drained for grazing. Some blanket bogs have been damaged by burning, including as we've seen to create better grouse habitat. Some have been dug up for fuel. Many lowland peatlands have been ploughed for farming and others mined to supply the gardening trade with peat-based composts. Even when damage was done decades ago the peat can continue to degrade, releasing carbon dioxide into the air as it gradually disappears.

It's not only blanket bogs and other peat soils that are leaking carbon. As we saw earlier, repeated ploughing and intensive cropping leads to sharp declines in organic matter in other kinds of soils. This too is contributing to carbon dioxide emissions. When it comes to farmland it was estimated that across the UK net emissions in 2005 amounted to about six and a half million tonnes of carbon dioxide (nearly double the annual emissions from all the households in Edinburgh, Cardiff and Leeds).

So it is that soils, especially those with a high proportion of organic material, and particularly peatlands, are important in determining the UK's overall carbon emissions. Above ground, too, large stocks of carbon are held in different natural habitats, with woodlands and forestry plantations holding about 150 million tonnes.

Seeing the carbon for the trees

The amount of carbon once held in Britain's forests was of course very much bigger than it is today, simply because there was in the past so much more of it. Whereas the country was once nearly all covered with trees, today only about a tenth of the UK is wooded, with only about a fifth of that tenth described as ancient woodland (that is, comprised of characteristic mixes of mainly native species and wooded since around 1600).

With reasonably natural woodland on just over 2 per cent of the country and another 8 per cent or so with different kinds of less natural tree cover, including single-species plantations of conifers, most of the carbon-storage capacity that was once in Britain's woodlands has been removed. And as we saw in the last chapter the soils that were underneath our native forests were subsequently stripped of much of their carbon through farming. At almost every turn we have pushed carbon out of land and ecosystems and into the air.

Sometimes it is easy to gain the impression that this kind of large-scale environmental change is a modern phenomenon. When it comes to Britain's forests, however, the trend toward the near-total disappearance of natural woodland goes back a long way. It is for example believed that by the time of the Roman invasion of Britain extensive tracts of natural woodland remained only in the Scottish Highland and in less accessible mountainous areas in Wales.

Fragments of more or less natural forest are even scarcer today. One that has made a big impression on me is the Abernethy Forest in the Scottish Highlands. I have been lucky enough to make several visits to this amazing place, and to be inspired by what is surely one of the finest examples of this forest type remaining in the British Isles.

It is an area of ancient Caledonian Forest dominated by Scots pine but with stands of juniper bushes, cherry, aspen, birch, rowan and alder. There are wet areas of bog and bog forest that support rare plants and insects. The woodland is quite open with the hummocky ground covered with cushions of moss and a luxuriant layer of heather, blaeberry and cowberry. In places there are dense thickets of regenerating pines. Abernethy Forest is all the more special because of its setting in a wider landscape with expanses of blanket bog, lakes and mountain ridges. It is a beautiful and memorable place with a rich variety of mammals and birds, including pine marten, red squirrel, capercaillie (a kind of giant-sized woodland grouse), osprey, crested tit and the only species of bird unique to the British Isles – the Scottish crossbill. With a great deal of luck (and even more patience) you might just see Britain's rare native wildcat, now at the edge of extinction.

The flora is also exceptional, with a suite of plants characteristic of Britain's boreal forest, including the delicate double-headed twinflower, wintergreen and creeping lady's tresses orchids. For some rare species, including a number of fungi, lichens and mosses, these woods are their main, and sometimes only, British stronghold.

Native pinewood like this was once the predominant forest type in the Highlands. A British microcosm of the coniferous expanse that still covers much of the high temperate to sub-Arctic areas of the Northern hemisphere, from Canada and Alaska through Siberia to Scandinavia – and Britain. Today, however, only little fragments of Britain's needle-clad cloak remain. There are about 84 patches left, covering a total of around 180 square kilometres. Abernethy is one of the best and also the largest.

By contrast with the richness of these native pinewoods, much of the land around Abernethy and across the Highlands is now covered with far simpler ecosystems. Claimed from forests for sheep rearing, deer stalking, grouse-shooting and for plantations of exotic coniferous trees, including sitka spruce, most of what is now the normal Scottish landscape we see today is very far from natural.

Here in the old pine forest though is an oasis of diversity and a web of connections that are a truer expression of how the land looks in this part of the world under more natural conditions. Although immediately less obvious than its role in sustaining wildlife and creating breathtaking landscapes, this remarkable ecosystem is, in common with other boreal forests, also a powerful carbon dioxide pump, removing that gas from the atmosphere, storing the carbon in trees and soils while helping to replenish oxygen levels.

While peatlands are the UK's biggest carbon store, woodlands are our most powerful carbon dioxide pumps. It is estimated that in 2006 a net total of about 15 million tonnes was removed from the atmosphere and stored by the UK's forestry plantations and semi-natural woodlands, equivalent to around 3 per cent of UK emissions. That is a very big slug of carbon, especially when you consider the area of solar panels or number of wind turbines that would be needed to have a similar beneficial effect.

With this in mind it might be tempting to believe that planting lots more new woodlands would be a good way to boost this service. While that would generally help, if the trees are planted in the right places (not on drained peat for example), bringing neglected woodlands into productive use would be better still.

Old woodlands that appear to have reached a stable and mature state are sometimes considered as less powerful carbon pumps than new plantings of fast-growing young trees. But most British woods have been cut over regularly for centuries, and so even the older ones that we call ancient woods are still likely to be in the carbon-accumulating stage, if allowed to recover. For example, about a third to a half of the native woods in England were cut during the Second World War and therefore have tree stems that are at most only about 75 years old. They will grow for centuries more, if they get the chance, in the process steadily removing carbon from the air.

The UK has been a helpful backer of international efforts to conserve and restore tropical forests so as to reduce the carbon emissions coming from them. While we have become used to assisting countries such as Indonesia, or Brazil and Peru, to slow down forest loss and in the process conserve wildlife as well as carbon stocks, perhaps we should recalibrate the effort to do more at home as well?

Maybe if we saw the Scottish wild cat as our very own 'Highland Tiger' and put effort into conserving and expanding its forests, we might not only show leadership in the global struggle to save forests and wildlife, but also demonstrate how carbon can be captured while pursuing other goals.

Blue carbon

The British Isles is not a huge archipelago but does have many natural assets of worldwide importance. Alongside its peatlands, lowland heaths, limestone pavements and other terrestrial habitats are internationally outstanding features on its 12,429-kilometre coastline, including saltmarshes, estuaries and dunes. These host different communities of animals and plants, including large numbers of shorebirds.

Growing up in an industrial suburb of Oxford my childhood fascination with wildlife was fed as much by books as by direct experience; this was especially so when it came to wildlife that I couldn't find within cycling range, which included the coast. So one of my most memorable childhood experiences was a first visit, aged twelve, to the Gower in South Wales.

Lying to the west of Swansea, this diverse peninsula hosts a wide range of different habitats, including oak woods that drip with lichens and mosses, flowery open grasslands, dense reed-beds and bogs. It also has varied coastline, with rocky cliffs, dunes and extensive mudflats. One place that caught my youthful (but urbanised) imagination was Llanrhidian Marsh. This is an area of muddy saltmarsh on the north Gower coast. An expanse of rough pasture dissected by hundreds of tidal creeks, it is home to thousands of shorebirds – shelduck, oystercatcher, redshank and curlew among them.

Grazed by sheep and cattle the grassy-topped marsh is in part separated from the sea by an impressive string of dunes known as Whiteford Burrows. Where the dunes and the marshes meet open mudflats is the point at which waves are victorious in the never-ending battle between sea and sediment. Sitting atop those dunes and looking out to sea it is humbling to ponder the dynamic interaction between water, wind, tide, sand and mud and how that eternal balancing act is now weighed by the warming of the seas and melting of ice caused by human activities.

With this wider picture in mind, we can begin to glimpse the full value of these ecosystems. For not only are the saltmarsh and dunes outstanding wildlife habitats, they are also further examples of natural carbon pumps, removing greenhouse gas from the atmosphere, turning it into organic carbon and accumulating that in their ever-changing fabric.

One estimate suggests that the average global rate of carbon accumulation in saltmarsh is hectare for hectare an order of magnitude bigger even than that of peatlands. The reason peatland is more important as a carbon store is because in the UK there is so much more of it. But even with its comparatively limited extent, saltmarsh is still a significant factor in the UK's overall carbon balance.

Lying where it does, however, in that busy place between land and sea, saltmarsh has been under relentless pressure for centuries, from port development, demand for farmland and because of urbanisation. For example in the 25 years between 1973 and 1998 the area of saltmarsh around the Thames estuary went down by about 1,000 hectares, that is a loss of about a quarter of what was there at the start of that period. A similar situation has emerged across the UK, as the more sheltered estuarine areas that favour saltmarsh formation have also been the places where we have farmed, established towns and developed industry and infrastructure.

Comparable pressures have affected many areas of coastal dunes across the UK. Whereas Whiteford Burrows is protected, others are not. About 25 kilometres away at Aberavon on the South Wales coast industrial and housing development has displaced a large area of dunes. Walking among the built-up areas, little bits of marram-grass habitat can be glimpsed between the houses and shops, tiny relics of what was once a substantial dune system.

Where they are still intact, though, dunes and saltmarshes are playing an important role in the process of keeping atmospheric carbon dioxide levels lower than they would otherwise be. In common with some other coastal ecosystems (such as the mangroves that fringe many tropical and subtropical seas), they are self-adjusting in relation to sea level and in the case of saltmarsh actually rely on taking carbon from the air in order to be able to survive sea level rise.

Staying just a little bit ahead in the struggle between sea and sediment requires new roots and plant growth that trap mud particles in relatively calm water during very high tides. This in turn buries the old stems and roots while creating the conditions for the new growth that lifts the level of the saltmarsh. Organic material being buried like this is the mechanism whereby saltmarshes remove carbon from the air.

One piece of research from scientists at the Universities of Virginia and Edinburgh suggests that with a faster rate of sea-level rise in response to climate change, saltmarshes could bury up to four times as much carbon as they are doing now. These ecosystems can of course survive only a certain pace of sea-level rise. If it goes up too quickly the marshes can't keep up and will become shallow water habitats, rather than low-lying land that is for most of the time just above the sea. If saltmarsh is overcome by fast-rising water it can no longer provide the same carbon-removal services that help maintain climatic stability.

It is important to note that not only are saltmarshes capturing carbon but also a range of other substances that are being locked up in their muddy fabric. These include nutrients escaping from farming and sewage works and pollutants released by the industrial installations that still line many of Britain's estuaries. For example four areas of recovering saltmarsh in the Blackwater Estuary in Essex with a total area of about 220 hectares were found to be sequestering and storing each year not only about 690 tonnes of carbon dioxide equivalent but also 15 tonnes of phosphorus. By reducing the pollution that reaches the open sea, saltmarshes are helping to boost the economic value of other aspects of our natural environment, including fisheries and the clean bathing waters that attract tourists.

The case of saltmarsh is but one of many practical examples that remind us how important it is to keep the pace of climate change as slow as possible. Releasing greenhouse gases at a level that cause them to go too far too fast might well mean that many of the jobs being done by nature, including in putting a brake on the warming by taking carbon dioxide out of the air, are shut down, thereby making things worse.

As to the scale of the job being done it is known that the rate of carbon sequestration in sand dunes and saltmarsh is high. Saltmarshes take in and hold between about two thirds of a tonne to over two tonnes per hectare per year. A typical area of sand-dune habitat takes in between about a half a tonne to one tonne of carbon per hectare per year. In addition to the annual uptake of carbon dioxide, there is also the far larger amount of about 1,000 tonnes of carbon dioxide equivalent *stored* in each hectare of tidal saltmarsh. That is a higher per hectare storage level than tropical forest, and thus a very valuable asset in the campaign to keep climate change to manageable proportions.

Considering that there are about 56,000 hectares of dune habitat in the UK and about 45,000 of saltmarsh, the total contribution of these systems to our overall national carbon balance is not massive, but using technology to achieve equivalent carbon benefits to those provided by our coastal systems would be very costly. It would be even more so to replicate the carbon benefits provided by our peatlands and forests. As debate continues on whether and how to bring forward carbon-capture and storage technologies that would take carbon emissions from power stations and other high-emitting infrastructure and store them in depleted gas fields and oil wells under the North Sea, nature is doing part of the job already, and at a very low cost.

Priming the carbon pumps

The work being done by nature in carbon capture is rarely recognised or valued. Even those of us who have long been concerned about rainforest destruction, and its carbon value, often have little appreciation of what's on our own doorstep. And so the depletion and degradation of natural carbon-capture systems continues. That this need not be the case is, however, amply demonstrated by many different initiatives.

One example is the lowland raised bogs at Thorne and Hatfield Moors in south Yorkshire. During the 1990s and 2000s, with colleagues at Friends of the Earth and other conservation groups, I tried to help put these remarkable places on a path toward being protected and restored. For decades these wetlands (covering a total of about 15 square kilometres) were being drained and cut for peat to be sold in garden centres. The result was a dark brown moonscape dissected by hundreds of ditches filled with water the colour of black tea.

This kind of peatland is rare compared with the blanket bogs that cover many upland areas, and only a tiny proportion of what we once had still remains. Although much of what little is left is heavily degraded, with the right treatment such ecosystems can begin to recover. Through a combination of publicity, product boycotts, lobbying, letter writing and legal actions, in 2001 we finally managed to convince Michael Meacher, then Minister of State for the Environment, to revoke the peat-cutting permission and to buy out the business doing it (the American company Scotts) with a compensation package that would get them off the land and clear the way for restoration of the bogs.

Today these remnants of the wetlands, that for thousands of years dominated the floodplain of the levels in the catchment of the River Humber, are designated as the Humberhead Peatlands National Nature Reserve. Home to some 5,000 species of animal and plant, not only are these recovering areas contributing to the conservation of some of our rarest wildlife, but much of the land is once more accumulating peat. These wet areas have thus gone from being a cause of climate change to being one small part of the solution – something of an achievement for a patch of the country that until recently many regarded as desolate wasteland.

Thorne and Hatfield Moors is, fortunately, far from being an isolated example. Across the UK conservationists are busy restoring formerly degrading peatlands. For example the Lancashire Wildlife Trust is taking forward work to restore parts of a once extensive lowland raised bog called Chat Moss. This is another area of raised bog and, like those on the Humberhead Levels, became heavily degraded. Elspeth Ingleby from Lancashire Wildlife Trust explained that it 'was once a ten-square-mile area of lowland raised bog but has been used for peat extraction over about the last three hundred years. Initially that was entirely by hand by local people using it for fuel and so on, and then with the building of the railway across the middle of Chat Moss, that led to much greater amounts of extraction as it was taken to Manchester.' Only three small fragments of the original area remain, and these are badly damaged. 'There's very little left – only half a metre of peat remaining from what was originally between ten and twelve metres across the whole site.' Even from this low point, however, there is hope. 'Our aims in the area are to bring each of our three sites back to restored mossland, to bring back the *Sphagnum* mosses, which absorb the carbon dioxide and lock that up, and then break down to form more peat, so it becomes a living moss once again.'

Ploughing drained peaty soil is also a big issue – but one that can also be addressed through changing how the land is used. One example of that can be seen at Woodwalton Fen in Cambridgeshire, where in 1912 Britain's first modern nature reserve was established. This precious patch of wetland had become an isolated island of ancient wetland in a sea of intensive farming. Before it too was drained, the land was bought by Charles Rothschild, a pioneering conservationist from the wealthy banking family, and protected for its wildlife.

Woodwalton Fen is still there today, and although some species have died out (such as the large copper butterfly that is now extinct in Britain), this area of fen still has much of the wildlife that Rothschild sought to save from oblivion a century ago. However, the reserve has become part of a far bigger vision, one that seeks to go beyond protecting what was left and to restore a larger body of nature. That vision is called the Great Fen Project, which is essentially a plan to reunite two now separate areas of semi-natural fenland (Woodwalton Fen and Holme Fen); this will be effected through the recreation of about 3,700 hectares of wetland habitats on part of the vast area of drained native fen which is now intensively farmed.

While some may bemoan the loss of agricultural land to nature (which being peat is in any event fast disappearing through turning into carbon dioxide), the Great Fen will bring a number of benefits. One is in relation to the job of carbon capture. Reversing the drainage and returning this area of the Fens to wetland will arrest the continuing degradation of the peat. Over the 80 or so years when the peat would have continued to disappear (with much of it expected to be gone long before then), each re-wetted hectare of the Great Fen will on average result in avoided emissions of 10,000 tonnes of carbon dioxide equivalent.

The Yorkshire Wildlife Trust is working hard to conserve peatlands in that county and the organisation's director, Rob Stoneman, sums up the value of that work in helping to meet the climate-change challenge thus: 'Eight per cent of all anthropogenic emissions to the atmosphere of carbon come from peatland. If you turn that on its head, working on three per cent of the global land surface can resolves eight per cent of the whole issue. And it's cheap!'

These and other schemes that are seeking to restore bogs and mires generally focus on re-wetting peat, so that it ceases contact with the air. As well as keeping water on the peat through blocking drainage channels, peatlands can be aided toward recovery by reducing the number of grazing animals, cutting the frequency of deliberate fires and taking steps to limit accidental ones.

In lowland areas, raising the water level in ditches, ending cultivation and ceasing to mine peat for horticulture are the main ways peat loss might be halted and new peat formation kick-started. The widespread use of alternatives to peat in horticulture is encouraging as well, although there is still a long way to go in getting it out of the vast amount of compost the country uses each year.

In terms of carbon, peat is a big deal. But the majority of our soils are mineral- rather than peat-based. They have less carbon content, but since they cover a much larger area they are still important in determining Britain's overall carbon balance. For example, England's grasslands store an estimated 686 million tonnes of carbon, while its arable soils hold a further 583 million tonnes. It is evidently worth taking action to keep this as high as possible, as part of a wider climate-change strategy. We saw some of the practices that could make a difference in this respect in the first chapter, including more thoughtful ploughing and measures to increase soil organic matter.

Woodland and carbon

There is also a lot of activity going on across the country to restore areas of ancient woodland. A key task for the RSPB and Scottish Natural Heritage in their work to look after the Abernethy Forest, visited earlier in this chapter, is not only to maintain it but also to expand it. This is being done in as natural a way as possible, and the forest is improving as a wildlife habitat at the same time. Through a strategy that among other things includes the control of deer (which eat young trees), the forest is spreading outside its core area. In time the newer areas of tree growth will start to look like the older parts do now. As it returns to a more natural state and occupies a larger area, so the carbon it captures and holds will increase.

There are many other examples of ancient woodland fragments being restored in England and Wales. One is at Bernwood Forest in Oxfordshire. Set aside by Anglo-Saxon kings for hunting, this once much larger 'forest' is today a mosaic of woodlands and meadows that was in the 1960s overplanted with stands of coniferous trees. It sounds almost unbelievable, but to prepare this ancient woodland site for the establishment of a commercial conifer plantation the forest was subject to aerial spraying with 2,4,5-T – a powerful herbicide that was one of the ingredients of the so-called Agent Orange used by the

US military during the Vietnam War. Agent Orange not only caused massive environmental damage there, but also led to a humanitarian catastrophe, seen not least in the terrible abnormalities caused in babies by the chemicals for decades after the spraying had ceased.

Such was the approach of the Forestry Commission to woodland management in Britain at that time. Fortunately, dormant seeds survived in the soil that lay beneath the dark plantations that replaced the oaks and other native trees, sitting there waiting for suitable conditions to return before germinating. In common with many other sites across the country this little fragment of woodland is now subject to the progressive removal of pines in the hope that the original mix of native woodland species will recover.

This work is worth doing not only for the long-term carbon benefits that can be achieved, but also of course for the wildlife. Bernwood Forest is one of the most important places in the country for butterflies, hosting several national rarities including purple emperor, white admiral and black hairstreak. The restoration is working and in decades to come Bernwood has a better chance of hanging on to at least some of the animals and plants with which pre-Norman inhabitants of this part of England would have been familiar.

Conservation projects like those that have improved Abernethy, Bernwood and hundreds of other woodlands have been and will remain an important part of how we'll get more carbon benefit from what are among Britain's most cherished and most characteristic habitats. In addition to this, however, is a range of climate-friendly commercial opportunities.

During 2012 I helped advise a group of business leaders that had been brought together by the British government as the so-called Ecosystems Markets Task Force (EMTF). The idea was to find ideas to help sustain natural systems while at the same time creating new business opportunities. One proposal related to the production of wood fuel. The EMTF calculated that the development of wood fuel from native woodlands could by 2020 generate over one billion pounds' worth of value to the UK economy and create and support over 15,000 jobs.

While there has been some controversy about feeding very substantial quantities of imported wood into large power stations as an alternative to coal, the idea from the EMTF was instead to supply local markets for small-scale use. An independent panel looking at the future of UK woodlands reached a similar conclusion.

It's not complicated, and in our house we see the benefits of what it means in practice. The very efficient Norwegian-made wood-burning stove that sits in our suburban sitting room is fuelled with logs cut from local woodland. Energy captured from sunshine in summers past is released to warm our home in winter. The carbon dioxide emitted from the combustion of the wood is compensated by new tree growth. As the woodland accumulates more carbon in the soil and plants, so the overall impact of this form of heat is relatively benign from a carbon dioxide point of view. If it's done right, wood fuel production can also benefit wildlife.

Wood fuel can help, but the climate-change benefit is all the better when wood is made into products that last a long time, such as structural building materials or quality furniture. Wood used for these purposes can take carbon out of circulation for centuries. Dr Nick Atkinson is a senior advisor with the conservation team at the Woodland Trust, a national charity committed to the protection and restoration of Britain's native woodlands. He gave me an impression of how much carbon dioxide can be removed by woods. 'The overall carbon benefits shake out to around five hundred tonnes of carbon dioxide per hectare over a hundred years, and that's the trees alone. There are also the soils in woodlands. They're doing the greatest bit of the carbon cycling that's going on.'

Like most conservationists he sees it as important to manage woodlands not only for carbon but to maximise a range of benefits all at once – including for wildlife and recreation. When it comes to the economic uses of woods and how best to align those with carbon capture he says it would probably be best if 'woodlands were being managed to produce timber of construction quality and firewood as a by-product of that.'

Having said this, some kinds of woodland management that are good for wildlife can also be good for fuel. 'In terms of purely firewood you've got the option of coppice management,' he says, 'which is great in terms of biodiversity because you end up with a range of habitat types.'

Coppice is a form of woodland management that is based on cutting young tree stems then allowing them to grow again from the stump, so that another wood crop can be taken later on. The process can be repeated indefinitely, with some British coppice woodlands having been subject to this kind of cropping for many centuries, in the process creating some of our most valuable wildlife habitats for birds, insects and plants.

Expanding wood fuel supply through the planting of new coppice and bringing more existing areas into traditional management could bring substantial conservation benefits, especially in those areas where there is little semi-natural woodland, such as where I live in Cambridgeshire. Across the eastern England region as a whole it is estimated that wood fuel production could be expanded threefold over current output to about half a million tonnes per year. That would feed a great many domestic wood-burners, among other things reducing our overall reliance on imported natural gas for heating.

David Ousby is the founder of a Community Interest Company called Cambridge Wood Works. His business runs a waste wood collection service that diverts wood away from landfill and toward reuse and recycling. What can't be recovered in this way, about 300 tonnes per year, is turned into Cambridge Hotlogs – a high quality fuel for household wood-burning stoves.

Ousby sees a big role for wood fuel going forward, although if we are to keep pace with rising demand he believes waste and existing woodlands will need to be augmented with supplies from new coppice. 'There is a lack of woodland in and around Cambridge. Nowhere near enough to supply the amount of fuel needed.' One idea he shared with me was about the potential for planting strips of native trees for coppice in intensively farmed arable areas. 'Ten metre-wide coppices of hornbeam, willow and hazel could be planted along suitable field margins, and would have other environmental and economic benefits. For example, when harvested willow leaf is an ideal cattle feed, reducing the amount and cost of hay and silage. Hornbeam can also be used for hardwood poles for fencing and hazel is ideal for natural fencing. All these species produce ideal wood fuel as well.'

Such linear woodlands wouldn't just provide carbon benefits but would create valuable wildlife corridors, boost pollinator populations, provide a haven for predatory insects that control pests, slow down water runoff and cut soil erosion. They'd also assist in a small way toward bolstering our energy security, as well as providing local employment.

About two thirds of Britain's wood demand is presently met through imports, and changing that proportion in favour of more home-grown resource would ensure we had stronger security of supply. If based mainly on expanding the cover of native broadleaf woodland, the landscape, conservation and recreational advantages would also be considerable. Various organisations are active in making this happen, including Grown in Britain, whose main aim is to expand domestically produced wood supply.

Saltmarshes and dunes

It's not only bogs and woods that could soak up carbon on a far larger scale if they were managed in a more optimal way. Different coastal habitats can help too, with saltmarshes having a particularly important potential role.

On top of the carbon benefits that accompany protection of this habitat type are those linked with wildlife conservation and fisheries productivity. Several fish species, such as sea bass (an important sporting and commercial fish), use coastal marshes to spawn. Then of course there is the work saltmarshes do in protecting life and property from inundation by the sea, as we saw for example in the new coastal wetlands at Freiston Shore in Lincolnshire. There are many other examples, including around the Essex coast and Bridgewater Bay in Somerset.

Dune systems too can be restored as well as conserved. The complex of coastal dunes on the Sefton coast of Lancashire and Merseyside is the largest in England and for years has been recognised as one of the most important landscapes for conservation in the British Isles. The dunes formed over many hundreds of years, with sand from the wide stretches of beach being blown inland and trapped by specialist coastal plants.

Damage inflicted on the dunes by farming and development caused wind-blown sand to swamp houses and fields. This is why restoration of the characteristic habitats has been a priority here for some time. Marram grass is particularly important for trapping sand and thus in stabilising the dunes, and as early as the 1600s officials called 'Hawslookers' were employed to administer fines to anyone caught cutting it for thatch. By the mid-1700s many local property leases imposed duties on tenants to plant this coarse tough grass.

In the early twentieth century continuing concerns about the effects of windblown sand led to the planting of pine trees. The same thing has happened in many other dune systems, including at Whiteford Burrows on the Gower and in parts of north Norfolk.

Efforts to stabilise and restore dunes are not only helping to maintain their carbon capture and storage function but are also helping to meet conservation goals. Many hundreds of species are found in the Sefton coast dunes, including some that are very rare, such as natterjack toads and sand lizards. Although not a natural dune habitat, the pine plantations are an important home for red squirrels.

All these restoration and maintenance activities require some level of investment, but when set against the costs of reducing carbon through other routes it doesn't look so expensive. If the co-benefits are taken into account, such as reduced flood risk arising from the dunes' sponge-like ability to absorb a lot of water, then the economics are better still. Looking forward and anticipating future climate-change policies, both in relation to how we'll cope with the impacts while cutting the emissions causing the problem and hanging onto our natural carbon pumps and stores makes a great deal of sense.

But while this context becomes ever more obvious, ministers are taking policy in what could prove to be a rather different direction. Many of the places I have mentioned in this short review of how Britain's nature provides carbon services – for example Abernethy Forest, Llanhridian Marsh and the Sefton coast dunes – are protected under European directives (on birds and natural habitats) that the UK government has publicly and repeatedly deemed harmful to economic growth. Despite the absence of evidence as to how these conservation agreements are actually detrimental to the UK economy, British ministers remain active in Europe to weaken the protection they provide.

No matter how much effort we put into reducing emissions from ecosystems, however, and no matter how effective the work we do to protect and enhance our natural carbon pumps and stores so that they take more out of the air and hold it, we'll need to do a whole lot more than this in the campaign to limit human impacts on the Earth's climate system. A big part of the plan must also include keeping carbon in the ground that has been captured by nature's carbon pumps up to hundreds of millions of years ago, including that held in ancient peat bogs far older than those that still clothe parts of Britain today. Our journey takes us next to see some of the ways in which we can do this, starting in the valleys from where hundreds of millions of tonnes of coal were mined to power Britain and its global empire.

Manifesto

- **Set out an ambitious new ecosystem** carbon strategy that identifies priority areas for action, including the target to restore 200,000 hectares of upland peat ecosystems in England, ending peat extraction and encouraging compost-based alternatives in horticulture (B&Q has already gone to peat-free growing media for the 140 million plants it sells each year).

- **Increase the staff and budget** of the Forestry Commission and direct it to provide more help to woodland owners so that they can produce more sustainable construction timber and wood fuel (for local consumption) while benefiting wildlife.

- **Direct the Environment Agency** to lead in developing a fresh approach toward integrated coastal management, including plans for the renaturalising of coastline, including the identification of priority areas for carbon capture and storage.

An offshore wind installation (top); and an artist's impression of the
Swansea Tidal Lagoon.

Chapter 7
Energy without end

ONE FIFTH – THE PROPORTION OF UK ELECTRICITY THAT WILL BE PRODUCED FROM CLEAN AND RENEWABLE SOURCES BY 2020

SEVENTY PER CENT – THE PROPORTION OF UK ELECTRICITY THAT COULD BE SUPPLIED FROM MARINE RENEWABLE ENERGY TECHNOLOGIES BY 2050

NUMBER TWO – THE POSITION OF GREAT BRITAIN IN THE WORLD LEAGUE TABLE OF COUNTRIES WITH THE LARGEST PER-PERSON CARBON DIOXIDE EMISSIONS SINCE 1850

During the summer of 1995 I was on a Welsh hillside searching for marsh fritillary butterflies. These gorgeous little insects, with reddish-brown wings chequered with darker brown, orange, yellow and white, were once quite widespread in damp grasslands across western Britain but because of habitat loss and changed farming practices have become one of our most threatened butterflies.

Known for their tendency to live in isolated colonies, I was searching for a group that had been found at place called Selar Farm, at the head of the Neath Valley in South Wales. Selar Farm itself was not a beautiful place, surrounded by regimented forestry plantations and the scars left by mining operations that took place there from the dawn of the industrial age. Its tussocky damp grassland was, however, ideal for the marsh fritillaries to carry on their precarious existence, not least due to the presence of the scabious plants upon which their furry black caterpillars dined. On that evening in early June I was delighted to find my special insects fluttering in a sheltered corner, where scattered oaks and stonewalls slowed the breeze.

I'd gone there not only because of my love of butterflies. I was at the time head of Friends of the Earth's wildlife campaign and had devised a plan to see if we could persuade the government to change the law so that key wildlife areas would be better protected. Selar Farm was one such place. Because of its rare insects and other natural features it had been officially designated as a Site of Special Scientific Interest (SSSI).

There are thousands of these places across the UK and hundreds of them were threatened, for example by housing developments, road building and pollution. Our campaign had very small resources and rather than embarking on what would have been the futile task of trying to avert the particular pressures facing each one of them, the plan was to focus awareness on a 'Magnificent Seven' imperilled places and use those examples to gain support for changes to legislation.

The reason I had selected Selar Farm as one was because a planning consent had been granted for the establishment of a huge opencast coal mine. If this went ahead it would obliterate the hillside and thus the butterflies' habitat. Direct action campaigners had begun to turn up to physically resist the bulldozers, but with the law on the side of the mining the butterflies had no chance. Far stronger than rules to protect them was government backing for the extraction of fossil fuel.

So it was that Selar Farm was turned into the Nant Helen Mine, one of the largest man-made holes in the British Isles – a vast crater about one and half kilometres across and a hundred and fifty metres deep. With a fleet of 300-tonne excavators and 100-tonne dumper trucks, the hillside would yield about five million tonnes of anthracite coal extracted from the thirteen coal seams that lay deep beneath its once lush grasslands.

The destruction of a rare butterfly's habitat was one thing, but as time has gone by the global implications of this industry have emerged as quite another. Sold on to electricity generating and other industry customers, the coal from that mine has since contributed significantly to the UK's emissions of climate-changing gases. Much of it has been burnt in RWE's Aberthaw power station near Barry. Coal combusted there and at other sites will ultimately lead to Nant Helen causing about nine million tonnes of carbon dioxide being released into the Earth's atmosphere.

King Coal

Coal is made of very old peat – and Welsh anthracite is comprised of plants that accumulated in wetlands during the Carboniferous era that ended about 299 million years ago. Like the peatlands that cover parts of upland Wales today, anthracite, its fossilised equivalent, contains a lot of carbon – and when we burn it, carbon taken out of circulation by photosynthesis more than 300 million years ago is once more released.

Energy supply is the biggest single source of carbon dioxide pollution worldwide and within that coal contributes the largest share. British coal burning is a significant factor as to why carbon dioxide levels are as high as they now are. So significant in fact is the British contribution that in the league table of historic per person emissions of carbon dioxide since 1850 Great Britain comes in at second place. Only Luxembourg made a bigger per person cumulative contribution over that period, mainly because of the big part that steel making has played in that tiny country's economy.

In the case of Great Britain the momentum of the industrial revolution created demand for more and more coal. It fuelled steam engines in factories, the boilers on ships and the locomotives that powered the fast-expanding rail network. It was used to make coke for use in the blast furnaces producing the iron that was key to the fabric of Victorian infrastructure. It provided a cheap source of domestic heating, both directly and through the byproduct of coal gas that came from coke manufacture (and which was also used for lighting).

Anthracite from South Wales was especially good for generating steam, and during the second half of the Victorian era demand soared. In 1899 output from South Wales reached 40 million tonnes (about one hundred times the annual 400,000 tonnes that has been produced each year from the gaping hole that is now below Selar Farm's former butterfly habitat). The peak in British coal output was reached in 1913 when 300 million tonnes were extracted from major deposits in South Wales, southern Scotland, Yorkshire, the northeast of England and the Midlands. So pivotal was South Wales, however, that the Coal Exchange in Cardiff set the international price for this key commodity. It was there in 1913 that the world's first-ever one-million pound business deal was struck.

From this high point began the decline. In 2008 Tower Colliery, the last deep mine in South Wales, finally closed. Notwithstanding a few opencast mines recently being given the go-ahead, the loss of a large coal-mining sector, along with the industries that were associated with it, has led some parts of the UK into a decades-long period of economic decline. As gas, nuclear and imported coal replaced domestic sources of fossil peat so jobs were lost and communities withered.

A visit to the South Wales valleys that once produced coal, where the past is visible in hillsides terraced with former mine workings, where spectacular viaducts, railways, bridges and other fragments of Victoria infrastructure point to the region's past industrial might and strategic importance, the changes that have taken place are starkly revealed.

In some villages shops and pubs are boarded up, areas of poor housing and dilapidated public buildings speak of social decline, where unemployment has afflicted several generations of young people and entire villages have died with the industry that once gave them life. It is possible to see here a little of the human side of what happens when a society changes its energy economy. The industrial decline evident in the South Wales valleys and in other parts of the UK is of course not the end of the story of how nature will meet our needs for power, heat and transport fuel. It instead marks the beginning of a transition toward that supplied by the Sun in real time, rather than captured by plants millions of years ago.

Bursting the bubble

There are several reasons why the shift from fossil sunshine to current sunshine – in wind, light and plants – makes sense. One is energy security: being able to rely on sources under our own control. Another is the long-term management of energy costs, as fossil energy supplies struggle to keep pace with rising global demand (as in the volatility in the price of gas in recent years). A third is the opportunity to lead in new technologies, generating jobs and profits in the process.

But above all these good reasons for making a fundamental shift in how we power our country comes the uncomfortable arithmetic of carbon dioxide and climate change. What quantity of greenhouse gas emissions can be released before major impacts result? A key threshold that governments, including our own, have said must not be breached is to go above two degrees of average global temperature increase compared with the time before the industrial revolution. This is widely regarded as a danger level beyond which it is believed our ability to cope with the consequences will be severely strained.

Researchers at the Potsdam Institute have calculated that in order to hold the risk of going past this two-degree limit to a 20 per cent chance, then the total emissions of carbon dioxide must be limited to 886 billion tonnes between 2000 and 2050. During the first decade of this century the world used up 321bn tonnes, leaving 565 for the forty years to 2050. That 565bn tonnes can thus be seen as a 'carbon budget', and like any budget it can be husbanded, carefully planned and spent wisely – or blown in a spree.

The compelling logic that flows from the idea of a carbon budget has been used by an organisation called Carbon Tracker to calculate how much of the fossil fuels we already know about across the world – the coal seams, gas deposits and oil fields – must remain in the ground. For the owners and investors in these industries it is not a happy conclusion, because the answer is that more than *two thirds* of it needs to stay where it is – right there in the rocks where it formed.

Some senior British government ministers and many commentators have become used to dismissing the UK's role in this task of reducing emissions because China and India have increased coal burning to the point where whatever we do will be overwhelmed by them. This point of view takes no account of our historical part in burning fossil fuels. It also fails to recognise our current role as a global financial centre for businesses based on these energy sources. Greenhouse gases generated from fossil fuel consumption taking place in the UK account for just 1.8 per cent of the world total. But London has assets based on coal, oil and gas listed on its stock exchange equivalent to nearly a fifth of the world's total remaining carbon budget. If the world is serious about low carbon development then most of the reserves listed in London (and elsewhere) will effectively become unburnable. That is a major strategic threat to London as a global financial centre, because assets that are presently valued in the hundreds of billions of pounds could suddenly crash, taking the share index down with them.

When there is an effective policy to limit greenhouse gas pollution then the true (and much lower) value will become apparent and fossil fuels will emerge as a financial 'bubble'. When that bursts, which it must do if we are to avoid going above two degrees of warming, the consequences are likely to be, in an economic sense, very unpleasant. Far better, some would say, to move investment in an orderly manner in advance of that, out of fossil fuels and into low-carbon alternatives.

It is of course possible that countries, including the UK, will give up on the two-degree target and decide that their interests are better served through policies and practices that permit continued high emissions. But what might that mean?

No one can predict the future, but what we can do is develop educated scenarios based on the best information we have to hand. To that end, in 2012 the World Bank published a report called *Turn Down the Heat – Why a 4°C Warmer World Must be Avoided*. This global champion for economic growth and development is not known for environmental alarmism or radical economics, yet it concluded that without action to cut emissions an average global temperature increase of four degrees could take place by 2060, and that it would rise to as much as six degrees later on, causing unprecedented heat waves, severe drought, and major floods in many regions, with serious impacts on the human economy.

This is but one of many recent authoritative warnings. Another came in 2012 from PricewaterhouseCoopers. This leading organisation that advises and audits many of the world's largest corporations told of the profound dangers that would come with the six degrees of warming that could plausibly occur this century. PwC's world-leading number crunchers concluded that in order to maintain economic growth without exceeding two degrees of warming the G20 countries need to reduce the carbon intensity of their economies at a rate of 6 per cent a year (that is to reduce the carbon emissions that come with each unit of gross domestic product by that amount). The five-year trend up to 2012 showed this ratio had improved by only 0.7 per cent, a figure exceeded by the G20 economic growth rate, therefore leading to continuing overall annual *increases* in emissions, not the pathway to the massive reductions needed. This put the world on the most extreme warming scenario set out by PwC's experts (6°C of warming) and on a pathway to use up the whole global carbon budget by 2034.

For anyone who doubts the extent to which this is now a mainstream economic issue, as well as a scientific one, then perhaps a few words from Christine Lagarde, the Managing Director of the International Monetary Fund (IMF), will help put things into perspective. In 2013 she described climate change in a speech to the World Economic Forum as 'by far the greatest economic challenge of the twenty-first century'. British government ministers like to quote the IMF when it brings favourable news about our economy. They'd be wise to repeat this too. The UK and all other countries face big challenges in relation to climate change and the sooner that is properly taken on board the better, not least because not doing so will court grave economic risk. That will in part be seen in the mounting costs arising from floods and storms but also in the growing vulnerability of our huge financial sector.

Mark Campanale is a financial analyst who's worked in the City of London for over twenty-five years and is one of the founders of Carbon Tracker. His work to understand how much of the fossil fuels we know about need to stay in the ground has led him to reach conclusions that financial organisations and government should take particular note of. Here's the crux of his analysis: 'The world's capital markets have by default assumed that all of the coal, oil and gas will get burnt and investors and bankers have thus deployed ordinary investors' savings into businesses that could lock us into a three degrees, even four degrees of warming.'

Campanale describes the trillions invested in fossil fuels as a 'massive misalignment between global financial markets and climate security' and sees investor backing to find yet more fossil energy reserves as unwise. 'We have shown how the top two hundred listed companies alone have reserves exceeding the carbon budget for 2°C and each year they are spending 674 billion dollars more to find and prove yet more coal, oil and gas reserves. We concluded that over the next ten years, if 2013's spending is replicated, that more than six trillion dollars could be at risk of being lost through fossil fuel assets becoming stranded.'

Looking at the increasingly serious impacts of extreme weather around the world (and the falling unit costs and increasing scale of renewable energy technologies), it seems inevitable that more and more countries will begin the journey toward low-carbon development. When they do, then the stranded fossil fuel assets will become a massive economic liability. Those placing faith in business as usual have evidently chosen not to see that change is already under way. The parallel with the 2008 financial crisis, and the unwillingness back then to appreciate mounting risk, is palpable.

Some get it, though. Take for example the giant insurer Storebrand. It recently pulled nineteen coal and tar sands (a very carbon-heavy fossil energy source) companies from its portfolio, citing carbon-bubble risk. The $800 billion Norwegian Petroleum Fund has halved its exposure to coal companies and in early 2014 the Norwegian Parliament voted in favour of completely divesting its vast national pension fund from coal. In July 2014 the British Medical Association pension fund did the same thing, with its members voting to take their investments out of fossil energy and put them into clean and renewable alternatives instead. Then in September 2014 the philanthropic Rockefeller Brothers Fund announced that it too was to divest its $860 million investment portfolio out of fossil energy companies. The heirs of John D. Rockefeller, who build his vast fortune with Standard Oil, signalled more powerfully than many others how the world is changing, with this symbolic oil-derived fund making its move to coincide with a major United Nations climate-change meeting. Hundreds of other funds are similarly divesting out of fossil fuels, or at least coal.

Campanale again: 'Investment banks have been running the numbers, with HSBC concluding that coal and oil company valuations are at risk in a future aligned with carbon budgets. Deutsche Bank, Citi Group and Goldman Sachs have all started questioning future fossil fuel demand citing the logic of unburnable carbon.'

While some defenders of the status quo present the process of decarbonisation as an impossible challenge that involves painful sacrifice, lost opportunity, insecurity and reduced competitiveness, it is increasingly clear that if we approach it in the right way it could be quite the opposite.

Queen Green

Richborough Power Station lies on the north Kent coast near to where in AD 43 legions of Roman soldiers landed, marking the end of the British Iron Age. Both the momentous local history and huge concrete testament to the age of fossil energy are reminders as to how dramatic change can occur very quickly. Between 1962 and 1971 Richborough burned over three million tonnes of coal. Although the tall cooling towers that loom over the flat countryside today produce no clouds of vapour, the cables that connect the mighty station to the national grid still carry electricity. It no longer comes from burning fossilised peat, however, but energy harvested from winds that drive the 100 huge turbines in the Thanet offshore wind farm that lies 11 km to the east in the Thames Estuary.

About 20 kilometres to the north, where the outer Thames Estuary meets the southern North Sea, an even more ambitious offshore wind farm recently came on grid. It's called the London Array and with 175 turbines each generating a maximum output of 3.6 million watts, it is the largest in the world, producing enough power for about a half a million homes.

It is a complex and challenging job getting this cutting-edge renewable power station operational. For four years specialist vessels laid the foundations and cables that connect the massive turbines to the National Grid. On a windy day in late 2011 I went out to see the Array during the latter part of its construction phase. By then most of the foundations were in place and the sturdy yellow-painted steel structures to which the turbines would be bolted were laid out in lines across the sea 650 metres apart with 1,000 metres between the rows. Two box-like substations that from a distance look like mini oil rigs were ready to collect the power from the turbines before taking the electricity ashore on massive cables to another substation by the Swale Estuary on the North Kent coast.

One of the construction vessels installing the turbine foundations was called *Adventure*. I saw this 450-feet-long and 14,000-tonne vessel at work. Struck by the bizarre sight of daylight beneath her hull, I watched as hydraulic jacks lifted the ship out of the water so as to create a stable platform for its massive hammer to drive foundations into the seabed in water up to 25 metres deep. This is heavy-duty, world-leading engineering, like our coal industry used to be. When complete the rows of turbines will be spread across a 100-square-kilometre area of sea. This kind of operation costs more than erecting wind turbines on land, although the higher up-front costs are partly offset by more constant and powerful wind.

Getting the London Array up and running has required huge investment, frontier technology, a complex supply chain and constellation of support services ranging from catering to shipping, all of which have created jobs. A walk around the harbour in Ramsgate from where the Array was being built and will be serviced during its operational life reveals a new vitality. The fact that in 2013 there were three times more people employed in Britain's wind-power industry than there were working in its coal sector says something as to how much things are changing.

For Ramsgate the arrival of offshore wind power, with its contractors, vessels, offices and support services, has provided a welcome boost to an otherwise slack local economy. During the peak of the construction phase about 500 people worked offshore with many more on land. When complete, about 90 full-time permanent employees will operate this new piece of national infrastructure, including highly skilled professionals. That is about the same number as employed at the Nant Helen mine. A gas-powered station of similar electricity output to the Array has a fraction of the number of jobs associated with it.

The construction of major offshore wind parks around the coast of the UK is thus a source of employment and investment in some of the most economically stressed parts of the country, including off the coasts of north Wales, East Anglia and the northeast of England. And there is scope for a lot more of it as long as government policy encourages investors to put their money into the industry.

The Array's towering turbines now loom over the choppy sea, the graceful 60-metre-long pale grey blades slicing through the air, driven on by a steady westerly breeze. Homes, shops and businesses are now powered in large part by these elegant machines.

Large areas of sea have been identified for a new round of offshore wind proposals and over time could make a huge contribution to national energy security. They could also bring a range of environmental benefits.

As we've already seen, large areas of seabed around the UK have been damaged (especially by trawl-fishing gear) and the expansion of offshore wind presents an opportunity to restore some of it. The foundations of sea-based turbines, and the rocky scour protection placed around their base, provide new wildlife habitats and where fishing is restricted for safety reasons the seabed effectively becomes a wildlife sanctuary. This can benefit both wildlife and help fish stocks, for example by providing areas where fish can spawn and feed, from there replenishing fishing grounds. Large aggregations of cod and pouting were found living around a Belgian offshore wind farm. Around a Dutch offshore array more porpoise clicks were recorded than outside – presumably because the animals were finding more to eat in there. Other studies have shown how the design of the concrete bases of the turbines can be made to attract more wildlife, such as crabs and shrimps that in turn are eaten by fish.

Professor Martin Attrill from Plymouth University, whom we heard in relation to his team's work at Lyme Bay, believes that several positive benefits could come with the expansion of marine renewable energy. 'Once they're up and running there is very little evidence to suggest that they're having any impact at all on anything, even seabird populations. It'll be very interesting to see what happens to these areas when left alone. For the first time in over a century, we relieve a bit of the southern North Sea from trawling. I think they're very beneficial.' As was his conclusion at Lyme Bay, he believes we could get more from our seas if we used them more cleverly. 'There's no reason why the marine protected area, the wind farm and the mussel farm can't be in exactly the same place'.

This is not to say that the range of effects that come with offshore wind are universally positive. Noise generated during construction can be harmful to marine mammals while some species of bird can be negatively affected as well (one phase of the London Array has been halted over concerns about bird-life). These and other potential downsides can, however, be managed, and if the process of construction is done carefully the overall environmental gains can go way beyond low-carbon power to embrace wildlife and fisheries benefits as well.

In 2013 onshore and offshore wind turbines generated nearly 8 per cent of the UK's power and reduced carbon dioxide emissions by about 11 million tonnes. By 2020 the wind should be providing one-fifth of the UK's electricity needs, with roughly half each coming from onshore and offshore.

Despite these evident and well-documented benefits, there have been vocal objections against wind power, mainly because to some eyes the turbines appear to be an ugly intrusion. While the way turbines look is a matter of opinion (and survey after survey confirms that a majority of people are in favour of more of them), the criticisms that have come in relation to, for example, this energy source adding hundreds of pounds each year to energy bills simply don't stand up. According to the British government's Department for Energy and Climate Change the amount we spend on supporting large-scale renewable energy is about £30 per household per year, with less than £19 of that going toward wind power.

Some of the more extreme charges of mass bird deaths caused by turbines need to be placed in context too. Yes there are collisions, but these need to be placed in perspective. For example research from the United States estimates that about 40,000 birds are killed each year in the US by turbines. That is a lot, but not as many as the 100 million killed by domestic cats and approximately 500 million that die after flying into buildings. This is not to say that impacts on wildlife are trivial. Planning decisions do need to take a precautionary approach in ensuring that turbines are located so as to minimise danger, especially in relation to rare or declining species.

Despite the hostility, the UK's wind, wave and tidal sector nearly trebled in size between 2010 and 2013, with investment increasing from £2.8 billion to £8.1 billion, generating in the process not only lots more clean power but thousands of jobs.

Sophie Bennett from Renewable UK, the industry body representing wind, wave and tidal power companies, told me about research she had been involved with on the number of jobs that would accompany expansion in these technologies. 'We looked at the period from 2013 to 2023 in three scenarios, low, medium and high. Right now the wind industry employs 18,500 people directly and 16,000 more indirectly. Looking forward to 2023 in our high scenario there is the potential for a further 70,000 employees.'

Most of this growth is expected in wind power, especially offshore. But this is not the only place where opportunities for expanding our supplies of renewable energy can be foreseen.

Solar's bright future

Travellers heading in or out of London through Blackfriars station will have noticed that for years it was subject to major construction works, including the station being extended onto the neighbouring Blackfriars Bridge. What commuters couldn't see from the train was what was going on outside, and how the upgrade would not only create a modern railway station but also a cutting-edge power station.

Over 6,000 square metres of solar photovoltaic panels were installed to create what is now the world's largest solar power system on a bridge. The panels provide half of the station's energy and each year save over 500 tonnes of carbon dioxide emissions. This technology basically converts daylight into electricity and worldwide it is one of the ways we can drastically reduce our reliance on fossil energy. The potential for this kind of solar power is huge, including in countries like the UK that lie quite far north and that have a lot of cloud.

One person who knows all this very well is Dr Jeremy Leggett. For years he worked as a leading petroleum geologist, making forensic examinations of rocks and the traces of ancient life forms left in them to seek out new sources of oil and gas. Then he read the findings coming from climate change scientists, realised the potentially catastrophic consequences that would come with burning more and more fossil fuels and decided to change course, first of all campaigning to change energy policies and then setting up his own company to do the job himself. The business he established is named Solar Century. Founded in the mid 1990s it was Britain's first solar power company and also the first to turn over £100 million. Since then it has won many awards and installed a lot of solar arrays across the country, including that one on Blackfriars Bridge.

Leggett explained to me how the bridge demonstrates the flexibility of solar power, deployed at scale even in deeply urban environments. He told me how a lot of effort went into making it an aesthetically pleasing design so that it added to the visual impact of the bridge, rather than detracting from it. 'If roofs were done like that all over the country with facades thrown in as well, how much electricity could we generate in this cloudy country of ours where the sun virtually never shines?' he asks. 'The answer,' he says, 'is near to what we currently consume.'

Despite these kinds of projections, and as is the case with wind energy, solar has a host of critics who say it won't work at scale. As time has gone on, however, real world events have proved them wrong, as combinations of renewable energy technologies working together begin to disrupt conventional energy systems in dramatic ways. Leggett speaks of how leading energy sector analysts now see renewable sources creating a 'death spiral' for the old coal- and gas-burning utility companies, especially when linked with the emergence of new storage technologies and electric vehicles. Technology development is also leading to falling costs, with Leggett among a growing number of energy experts pointing to how solar will soon become 'cheaper than all other potential forms of electricity generation'.

If you run a traditional utility or fossil fuel company then this is of course unwelcome news, and resistance from that quarter is one reason why we are still going slower that we could be in scaling up clean alternatives. For the mainstream energy companies it is really a question of 'change or die', says Leggett. And it is not only energy experts who've seen the writing on the wall. When it comes to the fate of our pensions and the threat posed by that fossil fuel bubble I described earlier, there are signs that people are making the links and adopting new strategies for future financial security.

Leggett told me how 'Just the other day, some of the young folk in the office came and said that they'd just signed this deal down in Hampshire for a two-megawatt solar farm.' He said that he was surprised that his colleagues were so pleased as the company had done many installations of that size. 'I hadn't made the connection. This one is a local community project. A two-million-pound deal and every single penny of that has been raised in the local community. They will be getting returns on that solar farm that are better than they could hope to get even on their pensions or any so-called savings accounts.'

Perhaps a situation similar to that which has emerged in Germany is beginning to unfold in Britain. There, nearly half of all renewable energy-generating capacity is owned by communities rather than big companies – a fundamental shift not only in the energy mix, but its ownership. And irrespective of who owns them it is worth noting that even when arrays like the one in Hampshire are sited in fields rather than on buildings they can be integrated with other priorities. Solarcentury has for example been working with the Bumblebee Conservation Trust to find ways of encouraging insects among the panels. Some solar farms have integrated sheep grazing, too.

The huge opportunity presented by solar power was for a short period backed up by strong UK government policies, notably the so-called 'feed in tariff' that encouraged the installation of small-scale renewable energy systems. In 2010 alone this succeeded in attracting 5 billion pounds' worth of investment and by 2011 25,000 people were employed in Britain's solar power companies with 39,000 more in their supply chains. A policy change brought in by the Coalition government in late 2011 reversed this period of rapid growth, however, leading to the sector contracting.

It is expected to begin a period of growth again soon, driven by among other things consistently falling costs – evidently not a claim that can be made by the gas industry or nuclear generators.

Harnessing waves and tides

In addition to the proven potential for solar and wind-power technologies there are others that in time could be especially important for the UK, if only they can be developed and scaled-up. One is wave power.

The string of storms that hit the UK during late 2013 and early 2014 brought chaos and misery for millions. They were at the same time a reminder of the huge power carried in waves, as we saw towering seas pulverising seawalls and seafront promenades. Even in normal conditions the rough water that so often surrounds the British Isles, especially to the north and west, is packed with energy. Whipped up by winds in turn generated by sunshine, it is another potentially vast source of clean power and pioneering, exportable technology, and for the UK, it is a source that has some significant advantages. Big waves are frequent in winter when power demand is higher. And waves also help to balance other renewable technologies, such as wind, thereby contributing to security of supply.

There have been a few attempts to develop systems to convert wave energy into electricity. One was installed on the Scottish Island of Islay, where in 2000 a company called Wavegen was the first company in the world to connect a commercial-scale wave-energy device to a grid. I had a look at this machine in its remote location on Islay's exposed west coast back in 2008. Along the rocky shoreline, with patches of pink-flowered thrift and parties of oystercatchers passing back and forth, the aptly named LIMPET (Land Installed Marine Powered Energy Transformer) nestled like its namesake molluscs among the rocks in its concrete shell, quietly generating electrical power. This technology works by using an enclosed wave-capture chamber to convert the force of the rising water column into compressed air. The rush of air, and the decompression that follows when the wave recedes again, is used to drive a turbine that generates electricity. In 2011 a larger 300-kilowatt machine was opened.

Another Scottish-based company developing wave-energy technology is Pelamis Wave Power. They are refining a different kind of device, one that is installed out on the sea rather than being attached to the coast. These machines are comprised of a series of tubes connected in a row with flexible joints between the sections. As the waves flex the tubes against one another compressed air is produced, which drives turbines. Their prototype device was the world's first commercial-scale wave -energy converter to generate electricity to a national grid from offshore waves. Following its successful testing, Pelamis secured an order from a Portuguese electricity utility to build a wave energy farm off the northwest coast of Portugal, at Aguçadoura, thereby demonstrating that a market for the technology already exists. The three-machine farm, with a capacity of 2.25 megawatts, is the first of its kind in the world. Such pioneering engineering projects already under way could be the start of something substantial. If policies and incentives can be ramped up to drive more investment for research and development, the UK could be a world leader in an emerging technology.

While all these renewable technologies are really in the end about harnessing energy ultimately derived from the sun, there is also power we can get from the moon – via tides. This has been on the agenda for some time, with a prominent debate having gone on for decades concerning the pros and cons of a large-scale barrage being built across the Severn Estuary. There is enough energy in the huge tides that go up and down the Severn's massive funnel-shaped water body twice every day to meet about 5 per cent of overall UK demand. However, many conservationists (me included) have been concerned at the damage such a scheme would cause to wildlife habitats of international importance, especially for various species of fishes and birds.

However, a new tidal scheme proposed in Wales shows how it is possible to minimise the risks of advancing one priority at the expense of another. This is the Swansea Bay Tidal Lagoon – a project that would involve building a sea wall to impound 11.5 square kilometres of Swansea Bay. Water flooding into the lagoon on the rising tide drives turbines, and when the tide goes back out the water is released to repeat the process. If it gets the necessary investment and approval for construction, it will generate 320 megawatts of power for an average of 12 hours per day (enough for 120,000 homes) and will do that for 120 years, saving about 216,000 tonnes of carbon dioxide emissions for each year it is in operation. Its power output would be as predictable as the tide timetables that can be bought in the local newsagents.

This is only a fraction of what could be generated from a barrage running right across the estuary from near Cardiff to Somerset, but there are many other sites where the huge tidal range in the Severn Estuary could generate power with lower environmental impact, including offshore.

Mark Shorrock, the Chief Executive of Tidal Lagoon Power, the company behind the new proposals, told me about his vision for a number of schemes like the Swansea one. 'We are making the case for a series of tidal lagoons, like a string of pearls around the country, each one delivering enough power to every single inhabitant in a fifteen-mile radius of where that lagoon is built.' Should his company succeed in building this scale of power generating infrastructure the contribution to the UK's electricity security would be considerable. Shorrock says that one thing that is lacking in going to scale is what he calls 'Victorian foresight'. The point is well made, especially considering how the coal industry that once thrived in the valleys around Swansea made this area such an economic powerhouse in the nineteenth century, and how with some vision it could be again.

On the environmental side of the equation the benefits go well beyond the carbon dioxide emissions the lagoon will displace. 'We'll create Sea Bass, Pollock, lobster and mussel habitat,' Shorrock says. There might also be benefits for the thousands of birds that use the internationally important mudflats of the Severn Estuary too. 'Low-tide feeders are actually going to have longer feeding times because when the tide is coming in you've still got low tide within the lagoon. It's changing the timing, but you still get the same exposure of mud for the same time more or less.'

The Swansea project will cost £900 million to complete with about half of that invested directly in the city region. Shorrock is convinced that if the economic evaluation of his plan were more comprehensive there'd be an even stronger case for it. 'No economist puts a value on the twenty kilometres of sea reef you've just created that's going to make your sea more fecund again, and nor does he put a value on the oyster breeding area and the creation of lobster habitat, and the fact that you're going to restore the biodiversity and health of Swansea Bay, which is pretty trashed at the moment.'

If Shorrock gets the go-ahead for the tidal lagoon at Swansea in early 2015 he says it could be generating reliable and clean power by 2018. This is of more than passing interest, as warnings are issued about a looming power gap in the UK caused by old coal-fired and nuclear power stations coming to the end of their lives. The Swansea scheme and the five others proposed could meet about 8 per cent of the UK's electricity demand. It would require an investment programme of about 35 billion pounds spread over the period from 2015 to 2027, and in so doing provide a safer haven for some of the money presently invested in that fossil fuel bubble discussed earlier. There would be an average of 36,000 jobs supported by the lagoon programme, with an estimated peak of 71,000 in 2021. That is quite an opportunity.

One encouraging development in October 2013 was the decision of the Prudential, a major insurance company, to invest in the scheme. Tidjane Thiam, Group Chief Executive, Prudential plc, described the scheme as 'emblematic' and asserted that 'such investments provide our customers with strong and sustainable returns, create good jobs and increase productivity and economic competitiveness.' He said his company was 'proud to play our part in the development of this world-leading renewable energy technology'. Clearly the shift of capital out of fossil energy and into clean alternatives is possible, as mainstream investors are beginning to show through practical decisions.

Another technology in which the UK could lead, and which is already being operated here, is tidal stream power. Instead of running turbines embedded in a barrage or lagoon wall, this works by using devices that look a little like wind turbines located in the stream of water that flows when tides turn. The moving water rotates the turbine and generates power.

One such machine is already in operation in the UK in the form of a 1.2 megawatt device installed in the fast-flowing waters that connect the narrow neck of Northern Ireland's Strangford Lough with the Irish Sea. Twice a day water flows through these narrows with immense force and it is from this that a huge twin rotor device has been generating electricity since 2008. Developed and installed by a company called Sea-Generation, it is another British renewable-energy first.

Is there a renewable ceiling?

On top of solar, wind, tidal and wave power, there is as we have already seen a role for biogas from anaerobic digestion and wood fuel from our native forests. Solar-heating technologies work well too, and so do heat pumps that take warmth from the ground. Liquid biofuels could have a role in the transport sector (including for aircraft), especially those made from wastes or algae rather than food crops.

Then there are a range of new technologies that make much more of the electricity we produce, whatever the source. These include LED lighting that slashes the power needed to light homes and work places and good old-fashioned insulation that reduces wasted heat.

There is, however, a loud chorus of dissent. One theme that is returned to time and again by those who believe renewables can't deliver is intermittency: with no sun at night and wind not blowing all the time, renewables can't meet our energy needs. This is a critical question in planning the proportion of renewable energy that it is feasible to generate as part of our national energy mix. I asked Dr Gordon Edge, the Director of Policy at the industry body Renewable UK (with which I work from time to time), what he thought the answer might be. 'As time has gone on, we have got better at managing intermittency,' he asserted. 'First we thought ten per cent, then twenty and it keeps moving on upwards. By 2020 we could be getting about 20 per cent of electricity from renewables and National Grid sees no particular problem with integrating that. Looking to 2030 we could be getting up to about 50 per cent. At that point we'd need to invest in more interconnection, so that we could be using Norwegian hydropower and Spanish solar, alongside smart grids and energy-storage technologies.'

It's not just industry experts who see this potential. The government's own Committee on Climate Change expects wind, wave and tidal power to contribute up to 44 per cent of the UK's electricity needs by 2030 – a proportion that is evidently doable given that other countries are already achieving this kind of renewable contribution. 'In 2013 Denmark got half of its electricity over the course of the year from wind. They are massively interconnected into the Norwegian and German systems and have some energy storage, but they did it,' says Edge. He points out there is no silver bullet technology that would enable this scale of contribution from renewables, but rather a combination of measures. In addition to international interconnections these include getting more storage into the system and the flexible use of a limited amount of gas (which produces much less carbon dioxide than coal).

More renewable sources of electricity could also help to decarbonise transport. Most major vehicle manufacturers are investing in the development of full electric models and it is a market that is set to grow as battery technology and recharging infrastructure improve. Batteries will be an important energy-storage technology alongside renewable energy and British research is contributing here, too. For example Jaguar Landrover, the UK's biggest vehicle manufacturer and one of the country's largest exporters, is putting serious resources into what could be revolutionary battery designs to power future electric models. Wind farms generating power at night while most drivers sleep is one way to capture and store renewable electricity, in the process reducing emissions from vehicles. In the future, power stored in millions of vehicle batteries connected to the grid could be a way of balancing supply and demand.

Critics of renewable energy sources say all this is too expensive compared with fossil alternatives. The apparent high cost of renewable energy is, however, an illusion born of a combination of how the wider impacts caused by fossil energy (such as the costs arising from climate change and health-damaging air pollution) are not reflected in the prices we pay for them and how globally public subsidies for fossil energy sources are five times bigger than those for renewables. If the subsidies were redirected and the real cost of fossil energy seen in prices then the economics of renewable energy would look very different.

When it comes to the real cost of oil, coal and gas, one way to make it more visible would be through a carbon tax, levied in direct proportion to how much carbon dioxide different fuels release. The UK has such a tool, called the 'carbon floor price'. When this measure was introduced the plan was for the cost of emitting carbon from power generation to rise each year, starting at £16 per tonne in 2013 and rising to £70 in 2030.

By sending a clear and consistent long-term signal to power generators the idea was among other things to close the price gap with lower-carbon sources, including renewables. That was until the Chancellor's 2014 Budget speech, when the long-term plan was suspended and the carbon floor price frozen at the 2014 level of £20 per tonne. While the reasoning behind this decision was to protect UK competitiveness by reducing industry costs, could it be that in the end the opposite will be the result? Think for a moment what investors in renewable energy might make of this kind of decision.

Siemens is a major investor in Britain's expanding renewable energy sector and is, for example, investing £160 million in new wind turbine facilities in Humberside and Yorkshire, in the process generating up to a thousand new jobs. A new turbine-manufacturing plant is to be built at Green Port Hull and a new blade factory in Paull in East Yorkshire. The investment is being made to meet demand from major new wind farms that the company expects to be built in the North Sea, including on the submerged Ice Age sediments that comprise the huge shallow water area of the Dogger Bank.

Siemens intends to supply an international consortium of power-generating companies who plan to install a total of up to 7.2 gigawatts of wind-power capacity out there (that's about the same as seven nuclear power stations). If it goes ahead at that scale, then investment of nearly £28 billion will be needed to make it happen, in the process generating around 17,000 jobs. This vast wind farm will be comprised of hundreds of monster turbines, generating six million, or even ten million, watts each.

Siemens' decision to invest in the UK was not taken lightly, and of course was made when there was still a policy to have a long-term rising carbon tax. As if it wasn't bad enough that the Chancellor signalled that the UK was more interested in business as usual, and burning more coal in our power stations for longer, a few days later the Prime Minister added his voice to the theme of going backwards on wind power by saying he wanted to introduce a ban on on-shore wind farms.

Talk about mixed messages. Companies who've already invested here are not happy about the uncertainty being created and those who could bring in tens of billions of pounds more are made nervous by anti-renewable power signals. We have immense renewable energy resources here in the UK. Couple that with our heritage of manufacturing expertise and offshore engineering know-how, and we are in a strong position to lead one of the world's fastest-growing industries.

Gas and nuclear

Despite the evidence showing how the UK could be powered almost wholly by renewables, loud voices call for other priorities – especially natural gas and nuclear power.

It is true that gas is generally a much cleaner fuel than coal, but it is unfortunately not clean enough. Continued reliance on large-scale gas-fired generation is not compatible with the scale of emissions cuts that must be achieved, a point that has also been made by the Committee on Climate Change. The large-scale exploitation of shale gas with hydraulic fracturing ('fracking'), a policy favoured by many government ministers and MPs, causes a larger-scale release of climate-changing emissions than conventional gas sources, including because of the extra energy needed to produce it.

We'll need gas in the short and medium term, but planning to expand our reliance on this fuel is not wise from a climate-change point of view, at least not without some clear policy as to how much of it should be produced and with that governed by a carbon tax. Responding to government enthusiasm for shale gas fracking, solar pioneer Jeremy Leggett told me how he thought British ministers should 'go on a field trip to Germany and look at how that country has gone from having no renewables to having twenty-eight per cent of their electricity from those sources.' He adds, 'Next year it'll be more.'

Electricity generated by currently available nuclear reactors is a relatively secure source and better than conventional natural gas in terms of its carbon dioxide emissions. It does, however, have a number of drawbacks that lead me to believe renewables is the better route. For a start it is expensive and will only become more so, not least as accidents in different parts of the world (most recently at Fukushima in Japan) drive higher safety standards. Renewables are generally more expensive now but the price for those will continue to fall, while the price of nuclear will most likely continue to go up.

Nuclear power stations also create radioactive wastes that are leaving huge bills to be paid by our descendants for centuries. It is also a highly centralised technology and therefore a concentrator of jobs. Renewables can by contrast be installed nearly everywhere and involve a range of skills to make them work, including the lower-skilled roles that so many of Britain's young people need to make a living. When it comes to Britain's new nuclear stations a lot of the jobs that come with them will not even be based here, because the plan is for new stations to be built and operated by state-owned French and Chinese organisations.

Far better is to make the most of nature's rich renewable sources of energy by channelling the many billions of pounds that we will all collectively pay out in the decades ahead through our energy bills in support of British jobs spread right around the country.

When one considers the speed at which renewable energy is gathering scale, delay could lead the UK to miss a major global opportunity. In 2012, in just one year, the amount of installed wind-power capacity in the world increased by nearly a fifth. The installed capacity of solar power nearly doubled, while the economies of scale that accompanied this rate of expansion caused the cost of solar panels to be cut in half.

During 2013 alone, China, our favourite country to hold up as the reason why we shouldn't act decisively to cut British emissions of greenhouse gases, installed about 12 billion watts (gigawatts) of solar generating capacity and some 16 gigawatts of wind power. Considering how a nuclear reactor generates about one gigawatt, China's clean energy programme represents a truly vast scaling-up, with installed wind-power capacity now at over 90 gigawatts and still rising fast. It is worth noting that while this revolution has gathered pace nuclear power has gone into global decline.

There is no doubt that big change is under way. Leggett believes we have reached a turning point. 'For someone like me who has been advocating this stuff for so long, I'm now living in a time when energy analysts are outdoing us in rhetoric about what's coming down the line.'

He wonders whether official reluctance will leave the UK behind in the emerging global energy revolution. 'There's definitely a sense of resistance, but the jury is still out on whether that will mess things up for renewables in the UK. On one level it doesn't matter that much, as globally you won't stop it because it's a global market, and the manufacturing will just continue to grow and the costs will continue to fall. Meanwhile, the costs of the old industry will go up and up.'

If the UK is to be a leader in a new clean industrial revolution, though, and to reap the many benefits that will come with it, our politicians need to send clear and consistent signals to companies and investors so that money flows into businesses that can deliver. During recent years such signals have been at best equivocal, and one effect of that has been not only to miss the scale of opportunity that could come with clean energy, but also risk driving investment out of the UK economy.

No wonder the UK has during recent years slipped down the league table of countries most attractive to investors in renewable energy. Ministers who say they're worried about the UK 'being left behind' in global growth industries and new technologies should note that among the countries now ahead of us in renewable energy are India and China.

So much for emerging growth markets, but what about the butterflies that fluttered on that Welsh hillside on that sunny afternoon back in 1995? Across the UK they continue to face a loss of habitat, but I fear this will soon be superseded by an even greater threat. In common with many other declining and rare animals and plants living in isolated colonies, their fate will more likely be determined by how much and how quickly the climate changes. No longer is it sufficient to set aside little natural areas and hope they'll safeguard the many benefits Britain's nature brings for us. Today we also need a much broader and more integrated approach. It is one that among other things must combine the protection of nature with managing risks to our financial sector while building new energy infrastructure. We know what is needed to deliver that, but the big question is whether we'll have the sense to adopt the solutions in time.

From the global context of climate change our journey now becomes more personal. For not only does nature offer the means for Britain to build and sustain a new and modern energy system, it can also be harnessed to protect and improve our health and wellbeing.

Manifesto

- **Adopt policies** that will enable at least 44 per cent of UK electricity demand to be met from renewable sources by 2030 and outline a strategy to meet 50 per cent of power needs from sensitively sited offshore wind and 20 per cent more from wave and tidal sources by 2050. Publish new guidelines to ensure that all of the above is achieved while taking every opportunity to protect and enhance wildlife.

- **Reaffirm the commitment** enshrined in the 2008 Climate Change Act to cut UK emissions of carbon dioxide by at least 80 per cent by 2050

- **Work with industry bodies** to set out an official strategy to ensure that the UK has the skills needed to achieve these targets, thereby maximising employment opportunities.

- **Implement an aggressive nationwide** energy efficiency programme, including for the retrofit of the UK's existing housing and commercial stock. This would not only help to achieve the above energy-supply targets but also save money for households and businesses.

The revived River Itchen brings huge health and recreational benefits
(top). Green elements are integral to the design of Alder Hey Children's
Hospital in Liverpool – and for proven medical reasons.

Chapter 8
Human habitat

UNDER 10 PER CENT – THE PROPORTION OF CHILDREN WHO REGULARLY PLAY IN NATURAL AREAS

£105.2 BILLION – ESTIMATED ANNUAL COST OF MENTAL ILLNESS IN ENGLAND IN 2009

245,000 – NUMBER OF PEOPLE PARTICIPATING EACH YEAR IN WHALE AND DOLPHIN TOURISM IN THE WEST OF SCOTLAND

My travels around Britain to find out what nature does for us involved a variety of transport modes. Trains, buses, long walks, car trips and cycle rides, including one that, bizarrely, was on an exercise bike. Pedalling away in a windowless room I was not in my preferred surroundings. But despite the lack of daylight and sky I was participating in an experiment to measure some of the things that happen to people when exercising in nature. As I turned the crank on the stationary machine a large video screen in front of my face simulated a ride through beautiful mixed woodland in springtime. I rode under a bright green canopy of new leaves, passed great ancient oaks and sunny glades. There was birdsong, blackcaps, song thrushes, robins and others, and also the aroma of pines.

The reason for doing this in the confined space was to replicate as far as possible a full sensory experience, sight, sound and smell, of exercising in nature, but under controlled conditions so that careful measurements might be made. Collecting the data were Mike Rogerson and John Wooller, PhD students who are part of the Green Exercise Research Team based in the University of Essex in Colchester.

Before getting on the bike I undertook a test for mental agility and also a more subjective assessment of my mood. After the bike ride through the woods I did the tests again, and compared with the situation before made improvements on both counts. I was typical in showing this effect. The researchers explained to me that they've found a consistent positive result arising from exercising in natural areas, compared with similar activity taken in urban environments.

Healthy nature, healthy people

While my simulated bike ride generated just one bit of data relating to me, the material gathered in that and many other experiments is part of a rich body of evidence demonstrating how people's wellbeing is enhanced through exposure to nature, and not only at the level of individuals but that of entire populations. There are, for example, studies showing that physical activity outdoors can have a more beneficial effect on physical and mental wellbeing than comparable activity indoors. Researchers found a correlation between taking exercise outdoors and self-reported improvements in mental well-being and greater feelings of revitalisation, increased energy and positive engagement, together with reductions in feelings of tension, confusion, anger and depression.

Another study reviewed ten UK data sets involving 1,252 participants and looked at the potential for improving self-esteem and mood through spending time in forests, on seashores, by rivers and on mountains and moorland. They found that exercise in nature led to positive short-term health outcomes. Every type of natural environment improved both self-esteem and mood, with proximity to water generating the greatest effect. The biggest overall gains were detected among the young, while the mentally ill exhibited some of the biggest improvements in self-esteem.

A larger study by Natural England, the official government regulator for nature conservation, collected data from 4,255 respondents to investigate feelings of 'restoration' recalled by individuals after visits to different natural environments. They found that time spent in more natural areas helped restore depleted emotional and cognitive resources and how this effect came from access to a variety of habitats. Coastal areas were found to provide the biggest benefit, although woodlands, hills and mountains showed comparable effects. This research also found a positive relationship with the amount of time spent in natural areas: the longer people stayed, the better they felt.

Research by the Japanese Forestry and Forest Products Research Institute looked at how people's exposure to forest and urban environments affected a range of physiological measures, including blood pressure, heart rate, and cortisol in saliva (a stress indicator). Their results were very clear, showing that 15-20 minutes in a natural environment, such as deciduous forest, were accompanied by significant reduction in blood pressure, pulse rate and cortisol levels, when compared with a similar amount of time in a busy city area. Subjective ratings, such as 'calm', 'comfortable' and 'refreshed', were reported as higher in forest conditions than those recorded in the city.

Several studies have compared the effects on mental wellbeing of just viewing photographic scenes of both nature and built environments. The results suggested that the natural scenes, especially those depicting water, had more positive effects on measures of emotional wellbeing, such as sadness and happiness, than viewing built environments.

There is also research material that suggests a connection between the psychological wellbeing of people and species richness. For example data from the USA indicates a significant positive association between life expectancy and exposure to wildlife. A study conducted in urban Australia found that both personal wellbeing and neighbourhood satisfaction rose in places with a greater diversity of wildlife species.

The expanding evidence base also has found that time spent in natural environments can improve self-discipline in children, help treat attention deficit hyperactivity disorder, reduce aggression and crime, promote better health in the elderly and delay the impact of Alzheimer's disease.

Few studies reveal what it is about nature that helps to improve human wellbeing in these ways (my bike ride in front of the screen is one that seeks to disentangle the role of sight, hearing and smell) but there are several components. One is the colours that predominate in outdoor settings – the greens and browns of vegetation, the white, blues and greys of skies and water. All these connect with us in what we see as beauty. Sound plays a part too, for example that of birdsong, running water, breaking waves or the wind in trees.

Since 2010 research has been conducted at Alder Hey Hospital in Liverpool to examine the beneficial effects that natural sounds, such as the weather and birdsong, can have on patients. Recordings of birdsong, rain and wind made by the hospital's child patients were played throughout the youngsters' wards to calm patients during injections, surgery and other stressful procedures. Laura Sillers, who helped to run the project, pointed out how: 'We have seen tangible benefits for patients in bringing the natural world into hospitals. We also installed it in the corridors and there have been numerous requests for the Bird

Song chorus to be reproduced on CD which patients then play at home.'

Smells, too, trigger associations with nature, for example that of wet grass, rotting leaves in a wood and flower scents. Humidity, temperature and light intensity all have effects on us as well. Dr Jo Barton supervises research at Essex into the effects of green space on people's health and says that it's the combination of sensory inputs that is important: 'The multi-sensory experience is what makes nature unique. There's also a lot of cognitive processing going on, and part of the green exercise benefit is that when we exercise in nature we work harder but perceive it to be easier, which has a positive effect on health and wellbeing.'

Whatever the combinations of triggers, a lot of science confirms that spending time in nature is very good for us. It's a conclusion that is not too surprising when you consider how we evolved in nature and only very recently became predominantly urban creatures. The momentum of our evolutionary past still attracts most of us toward nature, and yet we've largely designed our lives, cities and systems to isolate ourselves from it. Barton says that in order to maximise the benefits we gain from nature more effort is needed to encourage time outside, especially among the young: 'We need to encourage exposure to nature among children because the consequences of not being outside are having a negative impact on behaviour, cognitive function and health and wellbeing. We need to find ways of promoting that connection because otherwise these negative consequences will get worse.'

There is great potential for improving health outcomes among the elderly, too. Dr Rachel Bragg, one of Barton's research colleagues at Essex, has looked at how exposure to nature affects older people: 'Imagine you've been outdoorsy all of your life and suddenly you find yourself in a care home not going out anymore. It's horrendous. Dementia UK have done some work that shows the beneficial effects of green space on your mental state as you get older. And it's this aging population that is going to cost society a lot of money. Surely any positive interventions that reduce that cost are worth it?'

Considering how over 80 per cent of people living in the UK now dwell in towns and cities, these findings have huge implications. The Department of Health has reported some of the changes that lie behind our urbanised existence, including the fact that children now spend only 9 per cent of their time outdoors while for adults it is about 20 per cent. Less than 10 per cent of children ever play in natural areas, compared with 40 per cent of today's adults who did so as children. This dramatic and ongoing reduction in our collective direct experience of nature can be expected to have major implications in public health trends, reduced quality of life and lower life expectancy, and also in society's attitude toward the conservation of nature. Should more and more of us have less direct experience of nature, then demand to protect it will very likely also decline.

Lifestyles that involve more time indoors are often associated with reduced exercise, thereby increasing rates of obesity and type-two diabetes, while contributing to the higher incidence of depression and other psychological disorders.

It is perhaps logical, therefore, to expect that among people who live in towns and cities it would be those in neighbourhoods with more green spaces who'd tend to feel better, and this is indeed what was discovered through an investigation by Mathew White and his colleagues at the University of Exeter. Using data from over ten thousand people, and taking into account individual and regional factors, they found that on average both lower mental distress and higher wellbeing are linked with living in those urban areas that possess more green space.

These researchers showed how even small benefits to individuals can aggregate into large social impacts if urban green spaces touch many people (which of course most of them do). Importantly, they pointed out how efforts to protect and promote access to green spaces are often easier to implement than other population-level interventions aimed at improving public health (such as trying to influence diets or alcohol consumption). They also highlighted evidence that has come from a number of other studies indicating how the benefits of access to green space are disproportionately large among low-income groups and how interventions to increase access to it may help address social inequalities in relation to mental health.

The British government predicted in 2007 that by 2050 one in four under-16 year-olds will be obese, and that the total cost to UK society by then will reach about £45 billion per year (getting on for half of the current NHS budget). This is an eye-wateringly large sum, but the total cost of mental illness to British society is already far bigger even than that. For example, the wider economic cost of mental illness in England has been estimated at £105.2 billion each year, including direct costs of services, lost productivity at work, and reduced quality of life. If the United Nations' World Health Organisation is correct then by 2020 depression would be the second-most prevalent cause of ill-health worldwide.

In the light of research that confirms links between public health trends and exposure to nature, policy-makers have perhaps underestimated the potential for tackling some big emerging challenges. This is especially the case given how we are on average living longer and suffering less from infectious diseases and more from conditions linked with lifestyle and longevity. With less than half of all men and a third of all women believed to be sufficiently active to support good health, our collective vulnerability to cancers, heart disease, stroke and, mental and physical disability is expected to grow.

Dr Jo Barton sums up the opportunity to do better for health through nature like this: 'Many of the health trends that are afflicting society at the moment could in part be solved by getting people into nature. Having access to green space on your doorstep reduces health inequalities. It's that straightforward.' She emphasises the diversity of natural areas that can help do this. 'We have, albeit quite crudely, compared the different effects of different landscapes, from countryside to urban parks, from woodland to waterside, on people's wellbeing. All of them have positive effects, but the clear winner is waterside.'

Considering all this, and how a very high proportion of UK citizens live in cities, a key question for public health is how best to ensure people have access to areas rich in vegetation, rivers, lakes and shoreline. One obvious way to do that is to make such places available near to where lots of people live, and to get those places into a state that they want to use them.

This would not only be valuable in helping people feel better. Our innate yearning to be in contact with nature can also bring economic advantages.

Recreation

The fact that we get a lot out of nature is not only confirmed by the kind of research findings just described but is also in our willingness to pay for it. While many decision-makers can see the economic potential from new shopping centres and industrial parks, it seems that fewer have hitherto appreciated the contribution made by good-quality nature.

In addition to watching wildlife a great many people interact with nature via different sporting activities. The most popular is fishing with rod and line. With in excess of 2.6 million people regularly going fishing, it is the most widespread participant sport in the country. It can have an almost meditative effect, restoring calmness and feelings of wellbeing. And as well as helping millions of people feel better it brings a substantial positive economic contribution. Coarse fishermen (a term referring to the species of fish they catch rather than their manners) in England and Wales alone spend about three billion pounds per year, with more than 18,000 full-time and 5,500 part-time jobs dependent on their sport, including through the 2,700 or so shops selling fishing gear.

In some rural areas salmon and trout fishing is particularly important, especially in remote parts of Scotland where it is a key employer and generator of revenue for businesses. A 2007 study found that more than three quarters of salmon anglers visiting the Borders area of Scotland came from outside the country and spent on average £189 per day, compared with £56 by a non-fishing visitor. Across Scotland as a whole fresh-water fishing is estimated to support 2,800 jobs and generate £50 million in wages. All this is down to the desire of anglers to reap the benefits of being outside by water and interacting with nature.

Saltwater sport fishing is also a key economic sector in many parts of the UK. We saw earlier how at Lyme Bay the value of recreational fishing now runs into many millions of pounds, far exceeding the contribution that could come from alternative uses of the sea there, such as bottom-dredging for scallops. In that case the economic benefits that followed the restoration of the marine reef are very considerable, as are those linked with health and wellbeing. Similarly positive impacts can come from the restoration of freshwater habitats.

Many of our rivers have been turned into muddy drains, destroyed through the loss of their water sources or polluted with farm and industrial chemicals. While these abuses are usu-ally justified with different economic arguments (and excessive cost often cited as a reason why more can't be done to protect or restore them), rarely is this placed in the context of what has in an economic sense been lost, including the recreational and health benefits that once came with healthy rivers.

As we saw at the River Beane near Stevenage, where Charlie Bell told me how the decline of this once beautiful chalk stream brought an end to fishing, boating and swimming, the local people lost an important amenity for their wellbeing as the river was killed. The local economy has undoubtedly been harmed too.

Perhaps we'd be inclined to take more positive steps toward improving the state of natural features like the River Beane if it was clearer as to what could be gained as a result. We've already seen the substantial benefits in terms of cleaner water and flood protection that can come with restored water environments. On top of these, however, there are often a series of co-benefits linked with recreation, health and the economic value of water-based sports.

This has certainly been the case on the River Itchen, where we saw earlier the re-naturalising of a section of that river and its adjacent meadows improved not only the wildlife but also the fishing (and at the same time reduced flood risk). One indicator of the extent to which a more natural river increased the value for recreation is seen in how the price paid to the conservationists for the fishing rights increased fivefold after its transformation. And it's not only those who go fishing who benefit from the restored habitats in and around the Itchen for their wellbeing. Thousands of people live within walking distance of the river and the meadows can be accessed from a footpath that starts in Winchester.

There are lots of different ways to cut the economic figures that relate to the value nature provides through tourism and recreation, but the overall point is that good quality nature is not just nice to look at and great to be in, but also has tangible economic benefits that often come in tandem with what it does for our health. The great news is that it's not just the remaining far-flung wild areas of Britain where this is evident.

Reservoir ospreys

I mentioned earlier the potential for taking pressure off what were once some of our most beautiful but yet are now most threatened rivers, through the construction of water-supply reservoirs. At the same time as doing the vital job of improving water security, an especially important function as we enter a period of climate change, such infrastructure can be harnessed for other purposes.

Lying about twenty miles east of Leicester and right next to the town of Oakham, Rutland Water is the largest manmade water body in Western Europe. Construction of its dam, pumping stations, tunnels, pipelines and treatment works started in 1972, with the reservoir initially filled over three years between 1975 and 1978. While it is an artificial feature operated by Anglian Water as a commercial asset to provide public water supply, it teems with wildlife and is a magnet for visitors.

Carefully designed reservoirs topped up from rivers when they are overflowing from winter rains can be used to manage flood risk while at the same time taking pressure from wetlands and aquatic systems, including chalk rivers, enabling them to recover and in so doing enhance the range of economically valuable benefits. They can at the same time be hugely important recreational assets.

From the start Rutland Water was conceived with this in mind, with zones for wildlife conservation, fishing and water sports all reflected in the original design. It has been an outstanding success, not least seen in the fact that some 1.2 million people visit there each year. It has also been a success from the point of view of water security in the densely populated centre of England. It supplies more than half a million homes from water pumped in winter from the Rivers Welland and Nene.

Illustrative of the potential synergy between reservoir construction, the enhancement of the environment, water security and recreation is how each summer people flock to see breeding ospreys. These striking fish-eating predators lay eggs and rear young here, making Rutland the first English breeding home for these birds in 150 years.

A hide at the western end of the reservoir looks out over a swampy area of sedges and reed mace. Channels of shallow clear water divide up a wetland habitat that rings and jingles with the sound of several species of songbird, sedge warblers and reed buntings among them. Wildlife watchers can also get great views of the water voles that swim across the ditches and groom themselves on the banks. Across an expanse of open water is a platform with an osprey's nest on top. The birds arrive from West Africa in late March and stay around until September, when the adults and their new young depart south once more. This place is a magnet for visitors, with about 25,000 people stopping by here each year to see these magnificent birds.

Just before I arrived a group sitting in the hide had seen an osprey catch a fish right in front of them. That had certainly made their day. I settled down in the hope of seeing one myself. With me was Tim Mackrill of the Leicestershire and Rutland Wildlife Trust. His organisation has led the reintroduction of ospreys here and has seen the amazing effects that came in the birds' wake. We chatted about why people came to see the birds and he told me about the economic benefits that had come with more visits. 'People visiting us come from all over and so they stay for at least a day or two, a weekend. The local tourism industry has jumped on board with what we are doing.'

The ospreys attract attention further afield as well. 'Our webcam is watched by people all over the world, getting between 5,000 and 10,000 viewers every day. One lady was talking to us last year and she was from California. She watched the webcam every day and then, on a trip to Europe, actually came to see for herself the ospreys at Rutland Water.'

The webcams are also a resource for an educational project that is not only about conservation and nature but also cultural awareness. Involving over 100 schools from nine countries on the Rutland ospreys' migration route, pupils are not only learning about the interconnected nature of the world that we live in but also enabling direct interaction with other societies. Several schools are using the osprey project in this way so as to promote multicultural awareness.

Mackrill also told me how the birds had got under the skin of the local people, bringing a sense of specialness to where they lived. 'There's a letting agency called Osprey Lettings, the Osprey Restaurant at Barnsdale, the Rutland Ospreys netball team and a local school are calling their new nursery the "Little Ospreys". There are loads of examples of how they're being embraced by the local community.' The potential to get a lot of benefits from enhancing the natural values in new infrastructure has been well and truly proven at Rutland Water. The question is can we muster the imagination to do more of this? You'd hope so. As Mackrill puts it: 'What would you rather have? A reservoir that incorporates clean water, tourism, wildlife, education and recreation or an intensively managed agricultural landscape?' It's not of course a matter of either/or, we need both, but considering how we often struggle to achieve such good outcomes as those found at Rutland, and how building new reservoirs often attracts fierce opposition, his question is valid.

Ospreys are a case in point as to how reversing the fortunes of threatened wildlife can bring wider benefits to society. The first ones to breed in the UK for many years did so at Loch Garten on Speyside in Scotland back in 1959, as Dr Rob Lambert, Environmental Historian at the University of Nottingham, explains: 'We have a vibrant ecotourism industry here in Britain that was pioneered at Loch Garten with the ospreys. This is a site that's now been visited by more than two million people, sometimes ninety thousand in a single summer. The most watched bird nest in history anywhere on the planet.'

The recent return of ospreys to England and Wales has helped boost the total economic value of these birds to about £3.5 million, in the process helping to support local incomes and employment. It's not just ospreys of course. The RSPB has calculated the economic value brought by other birds, for example the magnificent white-tailed eagles that, following their reintroduction to Britain during the 1970s, once again breed on the Scottish Island of Mull. In 2006 the birds attracted about £1.6m in added value to the island's economy, through the money people spent while they were there. They've become even more popular since, bringing in up to £5 million to the island's economy each year and making a growing contribution to the total £300 million it is estimated that bird watching brings annually to the British economy.

Wild mammals, too, bring big value. During my visit to Argyll to search for beavers in the Knapdale Forest I spoke to several hotel owners and people running restaurants and pubs. All of them said the beavers had strengthened their businesses. In the early 2000s, whale and dolphin tourism in the west of Scotland attracted around 245,000 participants a year, spending £1.77 million on boat tickets alone. Total gross income into the regional economy that was attributable to the presence of these animals was believed to be around £7.8 million.

Wildlife tourism as a whole in Scotland is believed to generate more than a billion pounds of revenue and support over 7,000 jobs. Rural tourism more generally in England is believed to be worth around £14 billion per year. In 2002 it was estimated that over 17 per cent of all UK tourism trips involved nature or wildlife watching in some form or other, a proportion that has been steadily rising.

Rob Lambert puts it thus: 'We are a nation of wildlife watchers, whether it is seals, dolphins, birds, mammals, even butterfly chasing; add it all together and it's a strong, vibrant aspect of the domestic tourism industry. Certain charismatic creatures, largely predators, have captured the tourist gaze that used to fall on monuments and seaside piers and museums. That is hugely powerful for the conservation of those species – it pumps money into projects across Britain, but it's also extremely good for the tourists themselves who are contributing money and time and effort and making journeys often to remote locations to interact with species, contributing to rural economies in the process.'

No matter what the ultimate economic benefit, however, the whole thing relies on people seeking benefit for their well-being through interactions with wildlife. That's why they went to spend time in more natural areas – because it made them feel better. This is not a small thing, it is a fundamentally important aspect of our national health and one that it would be sensible to reflect in how we plan the future of our country.

A glimpse of what that future could be like is seen not only in those projects where nature was already integrated with infrastructure, but in new developments that it is proposed will soon be built. That new tidal-power lagoon that it's hoped will be installed at Swansea Bay is one example.

Like Rutland Water this huge and strategic undertaking is being designed from the start with people, conservation and its primary purpose of power generation all in mind at once. 'It will be a great amenity for people who live in Swansea. You will be able to get out on the ocean and people will be able to cycle around it. Triathlon bodies have asked if they can use it as their training centre and marathon organisers have asked if they can do a half marathon, or a marathon, on it,' says Tidal Lagoon's Chief Executive Mark Shorrock.

He added how the popularity of the structure for fishing was likely to be considerable: 'We've got local sea and pleasure angling associations, with over a thousand members, saying that this will be the best new fishing facility in the country.'

These and many other examples reveal how it is possible to enhance access to nature for human wellbeing at the same time as bringing economic advantages and developing essential infrastructure. And as Dr Jo Barton at Essex University suggested to me, we might expect considerable upsides through using nature to not only maintain health but to treat some illnesses. One organisation that has looked into the potential in some depth is Mind, the leading mental health charity.

Ecominds

Between 2009 and 2013, with the help of the Big Lottery Fund, Mind put in place 130 environmental projects as part of what they called the 'Ecominds' scheme. The idea was to help people with mental health problems get involved with nature-related activities so as to improve their confidence, self-esteem and physical and mental health. From regeneration projects in local parks to horticulture and from walking groups to farm-based activities, the ways that Ecominds sought to reconnect people with nature were diverse.

During the four-year period of the scheme some 12,000 people were assisted, with many finding benefits from the experiences they were offered. With mental health conditions it is difficult to generalise, considering their diversity and how they affect people in different ways, but case-study research by NEF Consulting gives an impression of what is possible.

Economist Olivier Vardakoulias wrote a report that included among others the case of a young woman who'd been hospitalised with mental health problems, was depressed, lacked confidence, and because of lack of exercise had become overweight and unfit. Ecominds helped her via a placement working on an organic farm called Growing Well, near Kendall in the Lake District. Among other things she learned to drive a tractor and earned a qualification for that. She enrolled on a horticulture course, gaining a qualification there too. Her health improved, she lost weight, becoming fitter than she had ever been, and she was discharged from mental health services. Having never had a job she began looking at how best to build a career in in farming. Two years after beginning her placement with Growing Well she secured an apprenticeship and a year later was in paid employment.

For a person regaining their health, this is a huge change, of incalculable value. For society too it is of great interest, not least because of the positive economic impact. Vardakoulias added up the prescription costs that were saved, the avoided medical consultation costs, the avoided costs of the community psychiatric nurse and avoided Jobseeker's Allowance to reach the conclusion that this young woman's recovery saved the state about £12,800 in one year. The cost of helping her in this way was just under £5,500 (including quite a lot of travel expenses that arose from the project being in a quite remote rural area).

Mind has seen many other people assisted by what the organisation has come to call 'ecotherapy'. Ken Ryan, their spokesperson, described the case of one person who'd made a particular impression on him: 'He suffered from post-traumatic stress disorder, he'd never considered nature as a possible solution and didn't even know that there was a nature reserve on his doorstep. He got introduced to it through his community mental health team and that reserve has been very important to him and has largely contributed to his recovery. He's coming out the other side now, his medication is being reduced and he says ecotherapy is imperative to managing his mental health.'

Ryan explained how Mind is not only looking at nature as a means to treat sickness, but as a way of sustaining health: 'We're very keen to see Public Health Commissioners using nature to prevent people from getting ill in the first place. We're piloting some ideas at the moment that take what we've learnt from ecotherapy and are looking at working with other providers to offer that service to public health teams.' He added how this is among other things one potentially important way to save the NHS a lot of money.

When you consider how each year about 4.6 million adults in the UK seek medical assistance for mental health difficulties (estimated to be about half the actual number of sufferers) it is easy to see how such numbers accompany the estimated economic cost of this worsening problem, to say nothing of the awful consequences for the individuals unlucky enough to experience it. Surely we should do more to leverage the learning found by Mind and others in helping people to rebuild their lives using nature-based therapies of different kinds.

The realisation that nature can bring advantages in the delivery of better health outcomes is an idea that's fortunately already becoming mainstream. Alder Hey Hospital in Liverpool, where pioneering work to bring nature to patients has achieved clear patient benefits, is stepping up how it does that and presents an example of what can happen when the dots are joined between nature and people's health.

In 2015 Alder Hey moved onto a new parkland site. Rebuilt next to the one-hundred-year-old hospital that was no longer capable of meeting the needs of the 270,000 patients who each year use its facilities, ecotherapy was at the heart of the upgrade. The modern building incorporates leading-edge natural design and when it's complete the old one next door will be demolished and turned into greenspace. Vicky Charnock, the hospital's Arts Coordinator, explained the thinking behind the rebuild: 'We recognise that nature brings massive health benefits to our patients. The whole new hospital is founded on those principles. We're the first of the kind not just in Britain, but as far as we're aware, anywhere in Europe. A hospital in a park, but also part of that park.'

The new building design has closely involved the children who are the hospital's patients. 'The building itself was inspired by a fifteen-year-old patient called Eleanor Brogan, who designed a flower shape for the new hospital,' Charnock explains. 'Between each of the petals we have green spaces, we have landscaped gardens, courtyards, play decks and spaces where our patients can be close to nature. Every room will have a view of the greenspace around them and even the roof is covered in grass to integrate it into the park. Our patients always said that they wanted to feel as if they were in the park and not in the hospital.'

Charnock and her colleagues believe that the new Alder Hey will revolutionise their ability to deliver world-class paediatric healthcare. 'We wait with much anticipation for how our new building and surrounding landscape will impact beneficially on the children and young people in our care,' she says.

In addition to the great wealth of material highlighting how bringing nature and people together can not only improve health but enhance sport, boost tourism and support jobs, there is a body of evidence pointing to how nature is linked with two other national obsessions.

House prices and crime

One organisation that has pioneered methods for the better use of green-space in ways that has led to elevated property prices and significantly less antisocial behaviour is the Green Estate Programme, initiated in Sheffield in 1999. It was one of the first projects to demonstrate the value of well managed natural-istic design in urban settings and the fact that it made such an impact in one of the most deprived neighbourhoods in the UK says a great deal about the potential for the kinds of methods it has used.

Green Estate was started through a partnership between Sheffield Wildlife Trust and Manor and Castle Development Trust, supported by Sheffield City Council, and was focused on the Manor and Castle Estates in Sheffield. This neighbourhood of around 10,000 homes formed one of the largest UK inter-war housing developments and had been labelled 'the worst estate in Britain'. An audit was undertaken of the area's green spaces and a vision of what could be done with them through the use of some regeneration funding was pieced together. Sue France, the Chief Executive, told me how they looked at 'every bit of verge, every park, every schoolground and shopping area and came up with this impossible to deliver, but imaginative project, which became the catalyst for change.'

It was not an easy place to talk about 'nature'. France said that the long-term impacts of multiple deprivations meant that key concerns focused around poor health, low employment levels, failing schools and how people in that postcode area struggled to get a job because the place had such a bad repu-tation. There was 'a real lack of community infrastructure; the outdoor environment was of the lowest quality and very high levels of antisocial behaviour undermined traditional invest-ment programmes and approaches. There was a huge reluctance to engage in short-term funded greenspace projects that were seen as imposing middle-class values, appeared do-goody, or talked about wildlife.'

When Green Estate was established the City Council had withdrawn much of its parks funding from the area, having come to regard spend on the natural and outdoor environment as a waste because antisocial behaviour was so rife. During an early phase of the work more than 70 burnt-out stolen cars were removed from one park alone.

Whilst it was a hostile and abused environment the area did have an abundance of green space. Green Estate identified about 280 hectares of green areas, including old watercourses, semi-natural grasslands, urban commons and even some ancient woodlands. Through engaging local people, and by employing them rather than bringing in contractors, the programme developed real momentum. Inspired by the impact this had almost immediately on antisocial behaviour and the pride people started to show in the projects, the pilots quickly grew into an area-wide programme. After two years, more than twenty people were employed.

Work got under way to regenerate the estates' other green and open spaces, building on local heritage and need but also introducing new approaches to naturalistic design. A major new approach was the introduction of meadow-like landscapes full of flowers. The effect on the people living there was quite transformative. 'When the flowers emerged, butterflies, bees and lots of other wildlife also arrived and suddenly people were out there enjoying their local spaces. What we really learnt was that to make nature appealing in this very stressful urban environment, where lives are challenging enough and you're overwhelmed with grot, you really have to bring in beauty, activities, and make sure there is appropriate management in place to make it appealing,' says France.

Working with the University of Sheffield the team tested new flower mixes for planting across wasteland and green deserts. This subsequently led to a business spin-off in the form of a subsidiary company called Pictorial Meadows, which sources, supplies and advises on the planting and maintenance of specialist flower seed mixes (it was their meadow design that helped the East London 2012 Olympic Park emerge as such an outstanding environmental success).

During its early years the Green Estate programme was heavily supported by British government and EU regeneration funds. By 2004 it was turning over around £1 million and employing about 50 people with the programme run through a community-owned social enterprise model. By 2007 it had become more or less self-sustaining.

Compared with some earlier efforts to change the culture of these areas, Green Estate made a big difference. A classic example is that of a playground built before it got going. The City Council had invested £400,000 on a multi-games area that lasted two weeks, before it was ripped to pieces and burnt down. The playgrounds that Green Estate put in during its first year have not only survived but are now being worn out through outdoor play. Maintenance costs have dropped dramatically, the value of the outdoor environment has risen steeply and most importantly people are now enjoying being outdoors.

At the beginning of the programme, the Manor and Castle Estates were almost 100 per cent council housing and the idea that people might choose to invest in the area was considered ridiculous. The few properties on the market were hard to sell and attracting rock-bottom prices – around £12-13,000. The first new housing, adjacent to new parks, was designed around ecological principles and soon on the market for around £90,000 each. The whole area is now popular, with developers looking to build out the remaining vacant development land.

With green spaces that were once seen as a liability now among the Estates' principal assets, France says that there is a 'slight feeling of smugness', not only because of how the theory and vision proved so powerful, but because the programme managed to get such dramatic results from quite modest resources. Now the plan is to make sure that well managed green infrastructure remains at the heart of urban living. 'We're working specifically with the new Sheffield Housing Development Company to look at embedding the learning that we've gained into new opportunities presented by house building,' she says.

Given the increase in wellbeing and the dramatic change in a deprived area of Sheffield it's perhaps not too surprising. After all, how many of us would choose to live amongst largely barren green and grey areas, when more beautiful naturalistic areas were an option, especially when that comes with a package that includes a significant change in social conditions, including less crime.

The attraction is all the bigger when the contribution that green areas in cities can make toward protecting people from pollution is revealed.

Cleaning our air

The government's 2007 Air Quality Strategy estimates that the damage to health caused by tiny particles floating in the air (coming from among other sources vehicle exhausts) costs the UK between £8.5 billion and £20.2 billion per year. This is very likely to be an underestimate, as it ignores the impact of poor health, taking account only of deaths.

The concentration of health-threatening particles in the air can be reduced through tougher engine standards, but also by leaves which intercept them before they are later washed to the ground by rainwater. Trees, shrubs and turf also filter air to absorb other pollutants including carbon monoxide, sulphur dioxide and nitrogen dioxide. According to the US Department of Agriculture a single tree can absorb up to eight pounds of air pollution every year, in the process, and among other things, lowering the risk of asthma attacks.

As the climate continues to change, urban vegetation can help us cope with some of the weather extremes that will come in its wake. In 2003 an extreme heatwave (at a level we can expect to be normal by the 2040s) killed tens of thousands of mostly elderly people across Europe – mainly in France but some in the UK too. Trees can play a part in helping us cope with this kind of challenge because leaves absorb and filter the sun's radiant energy, keeping things cooler in summer.

Then there is the role played by vegetation in helping to reduce flood risk. Cities and towns are covered with hard surfaces that prevent the absorption of water. At times of high rainfall this can result in large volumes of surface water runoff that can overwhelm drainage capacity, become concentrated and cause flooding.

Sustainable drainage systems (or SUDS for short) are an increasingly mainstream way of using more natural drainage patterns and processes through creating permeable surfaces (soil and vegetation) to filter, absorb and moderate water flow. Not only can this help reduce the risk of localised flooding, it can reduce pollution of watercourses and provide amenity benefits, including for wildlife.

An example of what's possible can be seen at Portbury Marine Village next to the River Severn near Bristol. Here the Persimmon Homes development, begun in 1999, has incorporated design to render its hundreds of house resilient, while at the same time improving the local environment for the people who live there. This is seen in an extensive wetland next to the housing where the water that runs from roofs and other hard surfaces is diverted into an area of damp grassland, ponds and ditches rich in wildlife.

Unlike some housing developments that have been given the go-ahead despite high flood risk, conditions were placed on the developers to build in an integrated environmental solution that protected homes while at the same time enhancing wildlife. I walked there on a spring day with Robin Maynard from the Avon Wildlife Trust. The conservation group has taken on responsibility for the management of the wetlands so that the wildlife is encouraged for maximum public benefit. The result is impressive. Barn owls nest there and a variety of wading birds can be seen around shallow ponds. Lesser whitethroats sing their repetitive refrain from hedges while the strident notes of Cetti's warblers ring out from their concealed positions in thickly overgrown ditches. As we spoke to people out for a stroll, it was clear how much the natural areas were appreciated by people living there. A school group of six- and seven-year-olds passed by. Their teacher pointed out a peregrine falcon to the children, the amazing predatory bird flying over the wetlands on the lookout for an unfortunate duck or pigeon to eat.

Maynard told me how during a period of recent heavy rains the wetlands had taken a vast amount of water from the housing development. The wetlands are not only the destination for a lot of water but also help protect homes from inundation from the nearby River Severn. 'It's part of the sea defences of the Severn,' says Maynard. 'It's a buffer and offers tide protection and would take a lot of water if it breached the sea wall. It wouldn't be going into that housing estate, or reaching the houses down at the bottom either.' It is a marvellous example of joined-up planning, of people, infrastructure and nature together.

While some new developments increasingly take account of these kinds of opportunities and incorporate vegetation and 'green infrastructure' into their plans, many still don't. This is not a rational way to shape the communities of the future.

Surely people would thank us for having a more thoughtful approach. This is especially the case when we know how urban trees, sustainable flood defences and wildlife habitats all help to sustain our welfare in multiple ways, and could do so much more with the right kind of planning. In a society that is increasingly socially divided, progress could also be made in closing some of the health inequalities that blight British society. This will require not only maintaining but also restoring, rebuilding and enhancing natural areas near to where people live, especially in those places that are most deprived.

Millions of people across the UK live in what are effectively urban deserts, in housing estates with virtually no gardens or sited a long way from natural areas. Even when there are green areas, many are, as was the case in the Sheffield estates, virtually lifeless. Most of the architects and planners who design such environments, and the ministers who pass the policies that guide what happens, live in leafy green suburbs or rural areas rich in green assets. Perhaps if they spent a bit more time living in the kinds of habitats they help create for others, then they'd design them differently, especially if, like many of the inhabitants, they'd be effectively trapped there on low incomes, without transport and unable to escape.

Healthy habitat

Nature is good for us on an everyday basis, and through enabling people to enjoy it a series of social and economic gains can be achieved. These are diverse and seen in improved health and wellbeing, reduced health-care costs and in the jobs and businesses that rely on people enjoying nature's benefits, such as fishing, visiting the seaside, hiking and nature watching.

As for what might be done differently, there are plenty of obvious conclusions. Perhaps the first is to recognise the vast opportunities we have to enhance existing green spaces, such as we saw in Sheffield, and by the River Itchen. There is a need to design beneficial green areas into new urban developments, including allotments where people can gain the psychological and physical health benefits that come with growing food. Then there is the scope we have to make the most of new infrastructure, as was the case at Rutland Water. As to how we pay for such gains, it seems to me that a similar situation prevails as we saw with farming, where public money is not being spent in ways that get best value for money. Whereas farm payments encourage or permit the kinds of practices that lead to soil loss, pollinator decline, water pollution and flooding, the way we prioritise health spending seems similarly lopsided, with far too much emphasis on treating sickness rather than promoting health.

What might be achieved through shifting a small proportion, say 1 per cent, of the National Health Service budget toward improving nature and supporting nature-based therapies? The evidence suggests it could be a considerable amount. A report by Dr William Bird, a former GP who has undertaken extensive research into the relationships between access to nature and people's health, estimated some of the benefits. Basing his figures on a series of specific indicative examples he calculated for instance that a park in Portsmouth could potentially save the economy £4.4 million, including £910,00 to the NHS. A three-kilometre footpath on the edge of Norwich could save us £1 million, including £210,00 for the NHS. When scaled up to the entire country, and backed by educational and other steps to encourage more people outside, then simple amenities like these would generate some very big financial numbers.

One indication of just how big can be seen in a study published by official conservation advisor Natural England.

In a 2009 report the organisation suggested that every pound spent on establishing healthy walking schemes could save the NHS £7.18 in avoided costs in treating conditions such as heart disease, stroke and type two diabetes. If every household in England had good access to green space, the report estimated that £2.1bn could be saved in avoided health-care costs. At a time of intense debate about the future affordability of our NHS, these findings should be of great interest to policy-makers and the public.

Conservation used to be about protecting habitats for wildlife. Today it is also increasingly about creating optimum habitats for us humans. No longer, therefore, can our approach be credibly predicated on the idea of us giving nature a home. Instead we need to see things as they truly are, and how look-ing after nature is very much about keeping *our home* beautiful and healthy. The Ancient Greek word for home, *Oikos*, gave rise to the prefix of 'eco' – the root of both the modern con-cepts of economy and ecology. Maybe we need to relearn an ancient truth, and to once more see that our home is in nature, and so is our economy – there is no other place that either could ever be.

As seen in the pages just past, some of the positive things that flow from good natural areas being close to people can be secured by voluntary groups, such as the Wildlife Trusts, National Trust, RSPB, Mind, Conservation Volunteers and outdoor activity groups, such as the angling societies, who do such important work in providing opportunities for people to reap the huge benefits that come with being outside in nature. This is not enough, however. Government also needs to step in to ensure that that the dots that connect nature, health and wellbeing are properly joined up.

Manifesto

- **Undertake an official review** to determine the best value-for-money strategy that could be achieved through shifting 1 per cent of the NHS budget to fund a new 'Nature for Health' policy, including the restoration of natural areas in and around urban centres (targeting the most deprived first) and through supporting nature-based treatment referrals from GPs and hospitals.

- **Amend educational legislation** so as to require that all children leave school with the knowledge, skills and motivation to care for and benefit from nature.

- **Instruct the official nature conservation agencies** to agree joint strategies with national and county tourism offices and nature-based sporting associations so as to promote contact with nature among a larger proportion of the UK's population, and to maximise the associated economic opportunities.

Sometimes we get it right – the National Forest.

Chapter 9
Imagine

60 PER CENT – PROPORTION OF BRITISH WILDLIFE SPECIES IN DECLINE

£100 BILLION – EXPENDITURE DURING THE NEXT 15 YEARS ON FARM SUBSIDIES, FLOOD DEFENCES AND MODERNISING BRITAIN'S WATER INFRASTRUCTURE

4.8 TIMES – RETURN ON INVESTMENT EXPECTED FROM THE NATIONAL FOREST BY 2100

As we approach the end of our journey to see what nature does for Britain it is time to take stock, to look forward and consider how we might do better as a country through the more thoughtful and intelligent stewardship of the incredible natural systems that sustain us. There are already many positive initiatives, working to maintain or restore Britain's nature, and in the process contributing to our economy and wellbeing. But the message that nature brings vital values that must be maintained in our national interest is one that still needs to be heard by most politicians.

In fact, some of the statements from government bring to mind a poster proudly displayed at my Church of England primary school in the 1960s. This was divided into two parts, one depicting 'before' and the other 'after'. On the 'before' side was a hillside covered with native deciduous woodland. Below it were little fields dissected by old hedges dotted with mature trees and between them a mosaic of meadows and crop fields.

A stream ran from the hill toward a more substantial river and where the two met was a wetland and swampy reed-bed. There were a few farm buildings, a little lane, and in the distance the edge of a town with a church spire. On the 'after' side of the poster the woodland was replaced by a dense stand of conifers. The hedges were gone and an expansive monoculture of cereals had replaced the mixed fields. The stream had been buried, the wetland drained and the river dredged, straightened and imprisoned behind high banks. The lane had become a dual carriageway and hundreds of new houses added to the edge of the town. I'd sit and stare at that poster, baffled by how the 'after' side of the poster was presented as the desirable situation. It puzzles me now, too. Those underlying assumptions that the natural features were regarded as having no value.

The poster was unfortunately far from some fanciful vision. It showed what was actually happening. During the 1960s the destruction of natural features across Britain occurred on a massive scale, pursued in the name of 'progress' and backed by a new consensus as to what technology could achieve for us. This was especially evident in farming, as tens of thousands of ponds, hedges and meadows were ploughed into the new monoculture landscapes, ancient woods were sprayed with defoliant and the land smothered in toxic chemicals, including DDT, then regarded as a wonder chemical and pest-control panacea. Nature had come to be seen as an impediment to be swept aside by technology and the drumbeat of 'growth'.

Despite all the evidence that shows how nature in multiple ways underpins our welfare and security, our politicians too often remain trapped in that 1960s mindset. Take for example Chancellor George Osborne's comments to the 2011 Conservative Party conference, where he argued that it was 'environmental laws and regulations' that were piling on costs. Shortly after, in his 2011 Autumn Budget Statement, he directed an attack on the EU Habitats Directive as a 'ridiculous cost on British business', claiming the regulations that implement this vital set of laws in the UK were burdening companies with 'endless social and environmental goals'. The implication was that we should do less to protect nature because it was too expensive and getting in the way of progress. Osborne was not effectively challenged by any of his fellow ministers, including the Prime Minister, who'd previously declared his aim to lead 'the greenest government ever'. Nor was he called out by the Opposition. Despite the absence of push-back, however, the Chancellor's claims were in any event shown to be without foundation. This fact was revealed by an official government review that followed his attack; it found that the only industry the EU's nature laws were slowing down was wind power (ironically, a sector that Osborne and many of his colleagues had themselves been seeking to restrict).

A more recent assessment from the EU Commission reveals that the Habitats Directive (and its sister legislation to protect birds) not only causes limited impact in halting developments but actually makes a huge economic contribution. It estimated that the cost of maintaining what is now the world's largest protected area network (covering 18 per cent of EU land area and a substantial part of the sea) is about 5.8 billion euros per year. That is a lot, but it is dwarfed by the annual 200 to 300 billion euros the same researchers estimated Europe gets back from that investment.

And yet the evidence of the huge returns we get from healthy nature seems to fall on deaf ears. In addition to repeated attacks on EU environmental laws, the Coalition government has made drastic reductions in the funding of official agencies charged with protecting our environment; it has amended the planning system so there is less protection for nature; it has reduced support for clean energy; and it has relaxed regulatory codes that could result in different kinds of environmental damage. The decision in 2014 to cut the level of farm subsidies directed toward environmental improvements came in a similar vein. A higher proportion of the £3 billion paid annually from our taxes to farmers will be allocated with no such conditions in place.

Considering the many ways in which nature sustains our wealth, health and security, these are more than oversights. They are costly strategic mistakes that are wasting money, destroying value and making us more vulnerable. So what is needed to put us in a better position?

Taking stock

An epic report commissioned by the British government called the *National Ecosystem Assessment* – basically a stock-take on the state of nature in the UK – provides a good start. It confirms how many of the UK's ecosystems are in a reduced or degraded state (including marine fisheries and some of the services provided by soils) and how this ecological decline has reduced the value we can get from them.

Although bringing in large part a negative message, this and other recent work does show how we have an opportunity to do much better – and there have been glimmers of light even in government. The *Natural Environment White Paper* of 2011, drawing on that same comprehensive piece of official science, concluded that in order to ensure future prosperity and well-being in the UK the steady and significant decline in natural assets needs to be halted and reversed. It proclaimed how we would 'be the first generation to leave the natural environment of England in a better state than it inherited'.

That was a key moment, because from then on it was official policy to not only conserve but to restore nature in order to protect the country's interests, and in so doing helped set the stage for a new era in conservation. But can it be done?

It is hard to overstate the scale of ecological damage that has taken place across the British Isles. As we have seen in earlier chapters nearly all of the flower-rich meadows have gone, only about 2 per cent of the ancient woodland is left, intact lowland raised bogs are pretty much entirely lost and very few natural rivers remain. Many of our soils are degraded, nutrients have built up in the environment and most ecosystems impacted by pesticides and herbicides.

Unsurprisingly, declines in most of the major wildlife groups – wildflowers, birds, mammals, insects and the rest – have gone along with this damage to our natural infrastructure. Many species are already extinct here and there have been huge reductions in the populations of wildlife once common across Britain. On a global scale, the UK continues to load the planet's atmosphere with climate-changing gases (albeit at a progressively lower level than was the case in the past).

Until quite recently society was quite happy to accept all this as the price of progress, with the clash between nature on the one hand and development, agricultural intensification and resource extraction on the other, generally presented in terms of striking some kind of 'balance'. In striking that balance, however, the tendency has been to place more weight on the 'development' side of the scales. That's why even when there is so little left, still the demands of 'progress' lead yet more to be sacrificed on the altar of 'growth'. The proposed new high-speed rail route from London to Birmingham and Manchester is a current case in point, which if built would go through dozens of remaining patches of ancient woods.

The fact that we have any reasonably natural habitat to go through at all, though, even if only little patches, is something of an achievement, and down to far-sighted conservationists who during the early twentieth century realised that Britain was about to lose pretty much all of its remaining semi-natural areas. Initiatives put in place then led in the post-war years to acts of Parliament that at least slowed the pace of loss. By then though the damage was widespread and for the most part it was only little areas of half-wild habitats that had been spared. The campaign I mentioned earlier, the one that started on the Welsh hillside where the marsh fritillary butterflies flew, and where the habitat was obliterated by the construction of that opencast coalmine, was very much a part of that rearguard action – seeking protection for the little bits that had somehow survived the centuries-long onslaught. Although the resultant Countryside and Rights of Way Act passed by Parliament in 2000 did in the end succeed in applying a little more pressure to the brakes, it was not enough, by a very long way.

By just how much it wasn't enough was starkly revealed in 2013 with the publication of a report called *The State of Nature*. This major piece of work, reviewing the mountain of information we have about population trends among over 3,000 British wildlife species for which we have reasonably good data, found how nearly 2,000 of them were in decline. Studies into the state of particular wildlife groups bring even more troubling news.

For example a review of British butterflies published in 2011 concluded that 72 per cent of species had decreased during the previous ten years. Research into extinction trends among plants across the UK found that in 16 counties a plant species was on average lost every other year. Another report published in 2012 estimated how the UK had 44 million fewer breeding birds compared with the late 1960s. Overall *The State of Nature* report found how one tenth of Britain's wildlife species are at risk of going extinct here. Considering that the animals and plants in question rely on the same ecosystems that we do, these findings should sound alarms at the highest levels.

Turning around this still worsening situation is not only about the practical measures needed to halt and reverse the still continuing decline; it also concerns the mindset we adopt in doing it. Part of that is about knowing that it's time to change gear, to shift up from protecting the last of our country's depleted natural systems to also waging a campaign for nature's recovery.

Little isolated patches of nature gradually lose their wildlife diversity, and as the climate changes this will happen more quickly. Tiny natural areas also often can't do all that we want in terms of, for example, flood-risk reduction, protecting water security, optimising carbon capture or providing recreational opportunities. That is not to say that the remnants are irrelevant. Far from it. They are more important than ever, but they must be reintegrated with the landscapes in which they sit.

Restoration at scale: the National Forest

There are not many examples of our government taking the lead in restoring natural values at scale. As we have seen in earlier pages most of the good examples are down to leadership from conservation charities and more recently a number of ecologically-aware companies. There are a few instances though of where official agencies have acted in ways that reveal the massive potential for public bodies to add value to our society through improving nature.

One can be seen in the English Midlands and a place that in 1990 was designated 'The National Forest'. This came from an idea to create a major new multi-purpose wooded area that was first discussed by the Countryside Commission back in 1987. The proposal was to make a bold and visionary intervention by planting millions of trees in a densely populated and economically important area, and in so doing bring benefits for people as well as nature.

The area chosen for this ambitious new scheme was based on linking up the ancient forests of Needwood and Charnwood, spanning an area of about 520 square kilometres across parts of Derbyshire, Leicestershire and Staffordshire, stretching from the outskirts of Leicester in the east to Burton upon Trent in the west. Early consultations revealed massive public support, with research pointing to the potential for considerable economic advantages. The project was to include the restoration of former mining areas, while farming businesses would benefit from rural diversification, including tourism. By blending commercial forestry with public recreation it would be designed to bring multiple values. It was visionary, disruptive and new. Despite some considerable risk that such a bold idea wouldn't work in practice the idea was nonetheless backed by ministers in Margaret Thatcher's Conservative government. The inaugural site was planted with the first new trees in 1990.

When the first plans were laid the land newly designated as 'forest' had only about 6 per cent tree cover and was among the least wooded parts of England. Through the work of the not-for-profit National Forest Company that has encouraged landowners to change how their land is used, millions of trees have since been (and are still being) planted. The proportion of forested land has more than tripled, to the point where in 2014 about a fifth of the National Forest was actually beneath a mosaic of trees and other habitats, equivalent in area to about 16,000 football pitches. The longer-term plan is to get that up to a third. On top of this are some 94 kilometres of new hedgerows and 150 wildlife ponds.

The fact that the National Forest land has a population of more than 200,000 says something important about the practical opportunities that exist for rebuilding nature where lots of us live. That 10 million more have their homes within 90 minutes' journey time, and that it is open to all for recreation of different kinds, adds to its message about the possibilities that come when official bodies plan for people *and* nature.

A review by an economic consultancy called Eftec looked at how the National Forest's environmental assets contributed economic advantages, and concluded that it scored well in a whole range of areas. It created jobs, added millions of pounds' worth of value to the region's economy through forestry and tourism, improved energy security (in the form of more abundant wood fuel), was an important resource for recreation and aided skills development. It also locked up a significant amount of carbon. Overall, between 1991 and 2010 the National Forest is estimated to have brought economic benefits 2.6 times bigger than the total of £89 million invested in making it happen. The same report projected that by 2100 the benefits will be 4.8 times the overall investment.

The wider economic upside brought by the National Forest is also seen in different kinds of private-sector investment, including house building. During the mid-1990s house prices in the area covered by the forest were behind regional averages, but they are now ahead, and there is provision for a further 9,000 homes to be built in the area out to 2030. So much for nature being a block to development. In this case it has attracted it. A 2014 report by Genecon found that more than one billion pounds had been invested in the forest and surrounding area since 2001. And it's not only money the forest has attracted, but also people, with a population of 250,000 expected by 2030 compared with the 196,000 who lived there at the start.

The National Forest is one among several examples of efforts to restore and enhance nature that we have come across in this book: the blanket bogs on Exmoor, areas of the Caledonian pine forests in Scotland, the marine habitats at Lyme Bay and the lowland peatlands of the Great Fen among them.

Between these and many other inspiring initiatives are signals of a big change in how we see the job of sustaining nature. No longer is the task concerned mainly with the protection of special places with rare species and habitats (vital though that remains), it is now also about how to restore natural systems, and not only for their own sake but because of the different ways that healthy nature improves conditions for people.

This new perspective can even be glimpsed in some intensively farmed landscapes, and seen in the protection and recreation of hedges, woods and small wetlands that are, among other things, helping maintain and restore populations of pollinators and natural pest predators. The RSPB's Hope Farm is one among several that are showing a new way. In our towns and cities the naturalisation of green spaces is helping to improve the quality of life for people, including through reduced flood risk and cleaner air.

Knowing what we know, it is ever more clear that if we wanted to we could restore nature at scale, if only we put in place and join up the policies needed to do it, and of course secure the money and allocate it to the best ends.

Intelligent subsidies

The recent period of spending cuts and financial restraint is a challenging backdrop against which to suggest additional funds for the restoration of nature. However, we could do a lot worse than looking at the money already being spent. Is it being allocated efficiently and in an optimum way?

Over the next fifteen years or so it is expected that in the order of £100 billion coming from bills and taxes will be spent in ways that affect land and water. This huge sum is comprised of a combination of farm payments (about £3 billion per year), flood-prevention works (about £600 million per year and rising), cleaning up the effects of flood events and investments by water companies to improve water quality. On top of this will be the money spent by charities and official government agencies whose job it is to conserve nature.

I mentioned earlier a piece of work that I was involved with on behalf of three major water companies (South West, Wessex and Severn Trent Water) via an economic consultancy called Indepen. We looked at how catchments might be better managed and investigated whether it would be possible to reduce overall costs by spending these different budgets in a more complementary manner, rather than at times in opposition to one another. For example, if the farm payments encouraged reduced soil loss and pesticide use, then that could help cut the need for expensive technology to improve water quality and engineered flood defences. By how much was an interesting question.

We made assumptions based on existing studies and practical experience, and using conservative extrapolations in looking at the potential for wider applicability. In costing projects to improve water quality based on better land use in catchments, rather than traditional engineering, we assumed savings of only about a fifth achieved by the projects we reviewed would be realised on a larger scale. But that still suggested net saving of £4 billion over fifteen years. And this didn't include a range of other benefits to society, including carbon capture, flood-risk reduction and wildlife conservation. When these – and enhancements important for recreation, including for bird watching, fishing and walking – are added, the case for taking this route to meeting our water-related needs becomes even more compelling.

Such an approach will of course require more integrated ways of making decisions, especially in how we look at food production. At the heart of that challenge is the need for us to re-examine our decades-long quest for inexpensive food that only appears 'cheap' because it excludes the many costs that come with for example soil damage and the impacts that it causes.

Considering all this it seems surprising that we don't have a more intelligent debate about the best use of farm subsidies, and how they might be deployed in relation to the billions we also plan to spend on flood defences and in keeping our water environment healthy. If spent more wisely, the billions our government allocates in farm payments could clearly bring social benefits beyond simply the illusion of 'cheap' food. After all, most of us who pay for farm subsidies and flood defences via our taxes also pay water bills. If one lot of costs is increasing others then that really is a waste of money and for politicians across the spectrum should be a priority area for action.

Dylan Bright from South West Water, whom we met earlier to talk about the company's Upstream Thinking initiative, puts it like this: 'The elephant in any room when you talk about this stuff is food production. At the moment we pay the minimal price for our food and wreck the landscape by doing it in the wrong place too intensively, partly driven by subsidies, and then we pay a fortune in tax to have the Environment Agency clean it all up. If we just paid upfront to have our food produced properly, we'd still have our food but you wouldn't have all these negative consequences.'

Clive Faulkner, following two decades of experience in taking forward the work we saw at Pumlumon in the Cambrian Mountains, takes a similar view: 'The source of the wealth in the uplands comes entirely from the natural environment, and if we invest in that, then wealth will be generated. The more we invest, the more wealth there will be. That wealth includes food production, but there's also floodwater reduction, improved water quality, carbon capture and wildlife.' He believes most farmers are willing to manage the land for these multiple benefits, but that they needed clearer financial signals to do so, including from official subsidies.

As we walked around Hope Farm in Cambridgeshire, where the RSPB has increased bird numbers threefold at the same time as the farm has made more money, I heard the same message from Martin Harper: 'If you want to deliver better value for taxpayers' money, first you design the scheme well and then make sure that there's the appropriate advisory service.' He makes the latter point on the back of years of research that shows how the schemes that have encouraged farmers to do better for the environment while sustaining food production have had good-quality advice as a core part of the package.

Harper adds how in recent years this been drastically cut back by the two bodies that were largely responsible for delivering it: 'We've lost the Farming and Wildlife Advisory Group, and Natural England is now half the size it was.' When you consider how over the next five years some £15 billion in tax-funded subsidies will be paid to farmers in England alone, the near-total absence of official advisory capacity represents a major lost opportunity. Most of the money will be allocated to individual farmers simply on the basis of how big their farm is: the more land you have, the bigger the handout.

EU rules allow some discretion to member states as to the proportion of the subsidies that can be directed toward environmental and other schemes, and this is where at the end of 2013 government across the UK (with the exception of Wales) decided to go into reverse. In England the proportion of farm subsidies spent on environmental improvements dropped from the maximum 15 per cent to 12 per cent of the total, Scotland went down to 9 per cent while Northern Ireland set the level at zero. Only in Wales did Ministers hold the proportion at the top level. This is not a good system. Surely *all* of this vast amount of money should be directed toward socially beneficial outcomes.

In seeking remedies I found that it is not only a matter of how much money is allocated toward environmental improvement, it is also a question of how farmers are incentivised. The point was particularly well made by Natural England's Andy Guy as we hiked across Dartmoor: 'Payments that come to farmers are still based on what we call "income forgone". Calculations are made as to how much money farmers are losing by participating in agri-environment schemes. So basically they're paid for farming less intensively. They're paid for not doing something, which is bizarre really when you stack it up against the strength of the argument for the economic benefit that can come from a well-managed and healthy environment.'

He foresees the opportunity for a better approach: 'You've got the benefit of open spaces for people to enjoy, you've got wildlife for people to enjoy, and all of the other benefits that come from that. You've got carbon storage, you've got clean water, you've got flood mitigation, and all of these stack up to a really positive economic argument, but we're still selling schemes to people on a negative footing – paying them not to do something.' He told me how he thought that a more positive way forward could reap bigger dividends. 'I think that farmers themselves would take greater pride and feel stronger in their dealings with the public if they were paid on that positive benefit. It would be clearer to the taxpayer what their money was being spent on, and what benefit they were getting from it. At the moment it just looks like a hand-out, and that's no good for anybody.'

He added how it wasn't really for lack of evidence that this situation prevails. 'The academic argument right across Europe in favour of natural services and their delivery and the economic benefits that come from those services, has been won.' Guy's points confirmed what I have believed for a long time and how the challenge is not about evidence or logic but more a matter of politics.

When it comes to the political choices that determine how farm subsidies are spent they fall in large part to the EU's member state governments. They collectively decide the rules of the Common Agricultural Policy (CAP) that determines a lot of what happens in individual countries. Over the next five years the EU budget will be about one trillion (a million million) Euros, about 40 per cent of which will go in farm payments.

With that kind of money a great deal of positive benefits can be achieved for society – if only the policies line up. Perhaps there'd be more public demand for that if there were more awareness that in the UK we pay out each year via our taxes about £400 per household on farm subsidies. Interest might be higher still if people realised that through official support for certain kinds of farming they'll be paying again, for example in more expensive water, flood control and insurance premiums.

This context has led an increasing number of experts to advocate the idea of 'payments for ecosystem services' – basically paying people who control land to enhance the jobs nature does for us so that society can get more value for money. Some of the examples that we saw on our tour of Britain – where water is being made cleaner, flood risk reduced, pollinator populations being restored and recreational opportunities increased – all demonstrate the huge potential for the concept.

The multi-billion-pound farm budget is one source of money to pay for that. Another is the vast amount that will in coming years be invested by infrastructure companies.

Green infrastructure

The official industry regulator established by government to protect the public interest in our water infrastructure is Ofwat. This body has a mandate to scrutinise the investment proposals of the water companies – and thus the size of your water bill.

If a water company wants to spend millions, or billions, of pounds for example building reservoirs or new wastewater-treatment plants, then its proposals have to first be approved by Ofwat. Considering what several companies have now proved is possible through working with land and nature, as well as engineering and technology, it is certainly time for this powerful body to come up with clear guidance as to how it expects companies to manage future bills through the kinds of 'green infrastructure' solutions that we saw earlier.

Better still would be a coordinated approach with the Environment Agency that is among many other things charged by government with the construction and maintenance of the country's flood defences. This is because some of the steps that water companies need to take in using ecosystems to secure clean water supplies more cheaply are often the same as those that will help reduce flood risk. The work we saw earlier on the mountains of mid-Wales, the hills of Dartmoor and the valleys of Wessex prove what is possible.

The same day I opened that phosphate recovery plant at Thames Water's Slough sewage treatment works I also attended a meeting in London with water industry figures to discuss how best to meet environmental goals at the same time as holding down household water bills. Among them were a group of investors – people who decide where to allocate capital so as to get a financial return. They were very clear that the main thing they needed was consistent, long-term policy from government and official regulators. If they have that then they are more likely to make the investments the country needs to upgrade its infrastructure. If they were presented with rules that would incentivise cost-effective investments in nature, then they'd do it.

The same goes for some of the companies who'd like to upgrade our crumbling power-generation sector. As we've seen the UK sits amid huge renewable energy resources, including sources of wind, wave and tidal energy that can be captured by technologies in which Britain could be a world leader. If government policy is clear, consistent and supportive, billions of pounds of investment could be attracted into the UK economy, generating jobs and driving technology to harness clean energy that will be available for us for ever, no matter what happens to the gas pipelines of Russia or the oil fields of the conflict-riven Middle East.

The contribution being made by clean and sustainable technology development is already considerable. A third of the UK's recent economic growth has been attributed to the proliferation of 'green' industries. What's more, the benefits of that have been spread throughout the UK, rather than being focused in the southeast of England, with significant hubs in Yorkshire and Humber and the northeast of England. The wind, wave and tidal industries saw the number of people directly employed grow by around 75 per cent between 2010 and 2013.

The job of restoring and harnessing nature so it can do more for us is not only about government policy, how public money is spent and policies set up to attract investment. Some businesses will act in any event, because it makes financial sense for them to do so, as we've seen in this book; companies such as Thatchers, who've encouraged bees to pollinate apple trees for their cider business, or the Brixham trawler businesses who've saved money while causing less harm to the seabed; or the water companies that are cutting their own costs and those of farmers through advising on how to use fewer inputs.

Even in the case of private-sector leadership, though, government can help to gain scale by backing positive practice in different ways. Ideas we came across on our journey around Britain include setting up a new fishing boat skipper training scheme, reinstating farmer advisory services and new guidance and rules to make it more attractive for water companies to work more with nature.

A Nature and Wellbeing Act

So why aren't we doing all of this? One fundamental problem is that policies that govern and guide the way different aspects of the environment, such as water and land, are managed are not well coordinated. This often means that decisions are based on considerations that are too narrow, with one challenge dealt with at a time, and in isolation from related questions. On the plus side, this represents a major opportunity to gain more value for the same or less cost. As we've seen, there is plenty of evidence and experience showing how restoring and enhancing nature can in many cases be a way of saving money.

Having established how we can restore and enhance nature and gain practical benefits through doing so, and having recognised that a lot of money is already available for doing that, if only existing budgets were spent more intelligently, what might it take to get some effective scaling-up of good practice? One thing that would help here is a new, visionary and forward-looking Act of Parliament that would embody the aim set out in that 2011 White Paper to be the first generation to leave nature in better shape than we found it.

A 'Nature and Wellbeing Act' could set out the overall inspirational aim 'to restore nature in a generation', and bring forward new duties to protect and restore the natural environment in ways that enhance its different values. It would be the framework within which the many ideas and examples we've seen could be harnessed. For example it could be based on the kinds of maps that South West Water is using to identify where it is most beneficial for them to invest their resources. Mapping areas of the country to see where the biggest benefits could be gained for, among other things, reducing flood risk, securing clean water supplies, rebuilding pollinator populations, enabling recreation, storing carbon and recreating wildlife habitats, would reveal the priority places to focus resources with money and expertise pooled for maximum benefit.

We should be confident that this is possible because as we saw earlier some of this is already happening, including at the National Forest where in a generation nature is in much better shape, in the process bringing positive returns on investment. The thing we are lacking, however, is a nationwide scaling-up, the means to stimulate bottom-up planning and coordination to achieve improvements at the level of entire landscapes.

That cannot be achieved through lots of piecemeal projects alone. It requires a strategy and frameworks to bring people together, so that their collective impact can be focused and made bigger than each working in isolation, or even in conflict with one another, as for example farmers and those trying to prevent flooding sometimes do at the moment. Action on a landscape level can only be achieved if lots of different actors come together. Again, we should trust this can be done, because it is being done elsewhere. I've helped with such work in Indonesia, where companies, campaign groups, conservation organisations, land-owners and government agencies are starting to work together so as to not only protect but to restore fragmented tropical rainforest landscapes, including for the conservation of key animals such as tigers. If that kind of work can be taken forward there, then believe me it can be done anywhere!

The idea of bigger-scale planning for the restoration of natural systems should apply to seascapes too, so that the large-scale development of offshore wind, tidal and wave power can be best integrated with fisheries recovery, nature conservation and recreation. The strategies and frameworks needed to do this require national laws, otherwise they won't work at the large scale needed – and that is why we need the people we elect to represent our interests to introduce that new Act of Parliament.

Getting a new Nature and Wellbeing Act passed by Parliament will require politicians to see the opportunity and have the courage and vision to go for it, bringing people with them as they take forward specific proposals. But the first thing we need is a shift in mindset, to go beyond seeing nature as a barrier to progress and instead as an essential set of assets that are vital for the country to prosper in the fast-changing world in which we find ourselves.

Part of what we need in doing this is a clearer understanding of the value of those natural assets. At the moment we don't have that and as a result nature continues to be damaged in order to sustain short-term indicators of economic 'growth'. This is a major problem and is why we need more realistic national economic assessments, with the losses accounted for as well as the apparent gains. If we are to know if we are making headway in 'restoring nature in a generation', then having a clearer sense of its overall health is essential.

The Natural Capital Committee that advises the UK Treasury on the value of nature has reached a similar conclusion. Its 2014 report said that government needed to find ways of integrating the value of natural capital into decision-making so as 'to enhance taxpayer's value for money and to generate net benefits for society'. They called on the Office for National Statistics to integrate the state of nature's assets (natural capital as it has become known) into national accounts and to find ways for other organisations to do the same, including businesses. It was a timely reminder as to how as a society we find it hard to manage what we can't measure and how the rather technical business of accountancy could have a profound impact on the future state of nature, if only we gathered the right kind of information and used it properly. Comparable to the gathering of information to achieve a Triple A credit rating for our economy, we could do a comparable job in aiming for a Triple A rating for our ecology too.

No matter how illustrative new measures might be, however, it is important to remember that many of our leaders in politics, business, economics and the media grew up with the message exemplified by that poster on the wall of my primary school classroom. It was a message born of a worldview conceived in a very real post-war economic and food security crisis – and it is now decades out of date.

Modern challenges are different and increasingly linked with nature's limited capacities being overwhelmed and its assets depleted. Alongside a debate about the role for new laws and more realistic economic assessments, maybe a new poster would help make the point?

On the 'before' side of the new poster we might see the countryside as it is now: tiny fragments of semi-natural areas in oceans of crop fields dissected by roads and urban areas. The conifer plantations seen on the 1960s vision are still there. The town is surrounded with sterile housing estates. The river is still in its canal-like form and has just been dredged. A large water-treatment plant strips soil and pesticides from the drinking water supply while a nearby sewage works discharges masses of nutrients to the environment. This side of the poster depicts where most of us live today.

For the 'after' panel, the one that depicts a possible future landscape, the overall message is not about turning the clock back, but looking forward and anticipating the challenges that lie ahead. The deciduous woodland has been restored from beneath the stand of conifers. The river has been renaturalised. People are boating, fishing and swimming there. A wetland has been created and people are visiting to see birds and other wildlife (like the beavers that made it). That same wetland is also helping to reduce flood risk by holding water in the environment. On the distant hills blanket bogs, woodlands and more natural grasslands have been restored. Because of this the dredging of the river below is no longer necessary. This has not only saved the money needed to do that job but also cut water bills. The road has fewer cars (many of which are electric) and there are now cycle paths threading through the woods and diverse farmland with abundant birds, insects and wild plants.

Clean energy sources are integrated into the landscape. Carefully sited wind turbines, rooftop solar panels, anaerobic digesters and wood fuel among them. Market gardens surround the town, supplying local fruit and vegetables. Healthy soils and robust wildlife populations mean that fewer chemicals are needed to produce great food.

We might like to also design a poster to describe the future of the seas around us. In the 'after' panel, fish stocks are recovering in a cleaner marine environment, offshore renewable energy installations are generating clean power, and wildlife is coming back, including marine mammals and seabirds. Coastal towns and villages are prospering, not least from tourism.

All an unachievable fantasy? I don't think so. It's certainly no less achievable than the 1960s version of progress that has trapped our imaginations for so long. If anything it is more do-able, not least because a growing body of evidence and experience tells us how it would be a more rational approach, from the perspectives of cost, security and wellbeing.

A price on nature

Recasting our relationship with nature in terms of essential economic dependencies, security and human welfare can create unease. Surely, some say, if we go down this route we are 'putting a price on nature' and in the process losing something very profound, and obscuring the real and intrinsic value of the natural world.

I see what they mean, but disagree. This is not least down to the distinction between the value of things and the price we put on them. Value is about understanding the real contribution of something, whereas price is about transactions that could lead to a change of ownership, including from the public realm to a private one. While it is quite right that not everything can be valued in purely financial terms (the beauty of a rare butterfly for example), some elements of the natural world increasingly can be. It is crucial to make the point that nature has a real worth in economic terms – and to correct that fundamental flaw in how we view the world will make a big positive difference in our efforts to conserve and rebuild natural systems.

Hopefully, this book has made the case for appreciating the multiple values that nature provides for people as well as the value it has in its own right. And as the economic and practical case for nature is established, the other values it embodies – the intrinsic, spiritual, aesthetic and scientific – are strengthened. Throughout our journey you will have noticed that nearly all of what I've said is needed to sustain and enhance nature's multiple values is down to new policies, laws and redirecting finance, all predicated on seeing things differently. It is not, as some would say, about 'privatising nature'.

Through his work in the uplands on mid-Wales, Clive Faulkner has reached a similar conclusion: 'People have felt that with the approach we have taken toward looking at the economic value of nature, we've been selling out, putting a price on nature, and that this is immoral in some way. Nature does have an intrinsic value, it completely does – and that's why I am here – but it also has an economic component, and if we can find ways to help support the intrinsic value we're on to a winner.'

After three decades of making the case for nature myself – in politics, with companies and in the public discourse – it seems to me that the main problem we have in reversing the still catastrophic impacts we are causing has little to do with people not liking nature or not believing it should be protected. The reason we still travel in a destructive direction has far more to do with an out-dated version of progress and the mythical tension that is portrayed between ecology and economy. We desperately need a new storyline.

From the work that I do advising companies, through programmes run by the University of Cambridge Institute for Sustainability Leadership and the advisory group Robertsbridge, I know very well that many major infrastructure companies, consumer goods enterprises, house builders, manufacturers and others are ready, willing and able to establish business practices to protect, enhance and harness nature in clean and sustainable ways, if only the rules and incentives are lined up so that it makes sense to them.

Public backing is in place too, as shown by countless local examples of pride and appreciation of enhanced nature and its economic and social benefits. We are led to believe that people don't prioritise nature in times of recession. But much of the polling that shapes the political narrative is based on crude questioning, with people asked to rank the economy, health, education, crime and 'the environment' as if they were entirely unrelated. That phrase 'the environment' is troublesome, too, suggesting remote places, abstract ideas and complex technical debate that has little to do with day-to-day concerns. If the question were couched in terms of wellbeing, water security, flood-risk management, food quality, energy security and all the rest, a rather different answer might emerge.

Shortcomings in polling methods notwithstanding, there is still evidence of massive public support for positive change.

Organisations ranging in size from influential national bodies like the Wildlife Trusts and RSPB through to much smaller local or specialist wildlife and nature groups, between them have about four and half million paid-up supporters – more than 10 per cent of the UK's adult population. That number nearly doubles if the membership of the National Trust is added in.

Some environmental organisations have continued to enjoy growth in support, too, even during the recent period of austerity. For example Friends of the Earth gained 30,000 more supporters in 2013 alone, largely as a result of people coming on board to back its campaign to protect bees (that number of new supporters, recruited in one year, is more than all the people who've *ever* been paid-up members of the United Kingdom Independence Party).

When the total membership of the three largest Westminster parties is added together it comes to just under 370,000, or less than 1 per cent of the UK electorate. With at least ten times as many adults signed up to a more rational approach toward the conservation of the natural world, then it seems our politicians are missing a very important opportunity to engage voters. A new Nature and Wellbeing Act would be one way to exploit it.

So it is that an increasing number of businesses are on board, the public is willing, the science is clear and examples of what can be achieved with thought, collaboration and joined-up thinking there to inspire confidence as to what is practically possible. All this and more confirms that there's space for a visionary new approach.

It is time to step up for nature, to reverse the centuries-long degradation of the natural systems that sustain Britain and instead to invest in their recovery, not only because nature is beautiful and must be saved for its own sake, but also because its protection is essential for our health, wealth and security.

Manifesto

- **Pass a Nature and Wellbeing Act** with the overall aim of 'restoring nature in a generation'. A core part of this new law will require local authorities to prepare maps of current and future ecological networks to deliver water, flood control, carbon, pollination, tourism and recreational benefits, thereby taking forward many of the earlier manifesto priorities. Such maps would among other things guide future farm payments.

- **Set up a ministerial group** between government departments to establish ways to make the best use of all money being invested in land and nature – including farm payments, conservation budgets, water-company investments and flood-defence spending, so that all achieve maximum public benefit.

- **Use British influence in the EU** to press for new rules requiring all farm payments (not just a small proportion) to be optimised to achieve multiple benefits in relation to soil protection, flood-risk reduction, wildlife conservation (including pollinators and pest predators), improved water quality, recreation and carbon capture.

Acknowledgements

In writing *What Nature Does For Britain* I have been assisted by a great many people who have been generous with their ideas, creativity, support and time. Two individuals have been especially pivotal. One is my researcher Lucy McRobert, who was relentless in her pursuit of material and superb in her determination and ability to process masses of facts into meaningful summaries. She also set up many of the visits to the places I've written about and organised the great body of information we accumulated in the process. The other is Stephanie Hilbourne OBE, Director of the Royal Society of Wildlife Trusts (RSWT). Steph inspired me to write this book in the first place and her team took the step of hiring Lucy to work as my researcher.

Work to understand the practical benefits provided by nature to the UK has been a fast-expanding field that has recently become rich with insight and data. It has been my privilege to enjoy the help of many of the experts involved with this and to have the benefits of their experience and conclusions reflected in what I have written.

On what our soils do for Britain, I hope readers will find this book all the better through the ideas and insights that came from Dr Arnie Rainbow of Vital Earth; Dr Simon Mortimer, School of Agriculture, Policy and Development, University of Reading; Dr Erin Gill, environmental journalist and historian;

Professor Jane Rickson of Cranfield University; Professor John Boardman, Emeritus Fellow: Geomorphology and Land Degradation, Environmental Change Institute, University of Oxford; Professor David Manning, Professor of Soil Science, School of Civil Engineering and Geosciences, Newcastle University; Professor David Powlson, Visiting Professor in Soil Science, Department of Geography and Environmental Science, University of Reading; Professor Stephen Nortcliff, Emeritus Professor of Soil Science, University of Reading; Lee Dobinson and Neil Hunter of Generation Energy; farmers Iain Tolhurst and James Chamberlain; Simon Evans of Thames Water and Baroness Sue Miller of the Parliamentary Agroecology Group.

On the role played by pollinators and natural pest predators I am grateful for advice and insights from: Professor Jeff Ollerton at the University of Northampton; Dr Tom Breeze, Research Fellow at the University of Reading; Rob Saunders from Ribena; Martin Thatcher, managing director at Thatchers Cider Company; Dr Michelle Fountain, Research Leader in Entomology, East Malling Research; Martin Harper, the RSPB's Conservation Director (who also provided comment and advice on wider matters of agricultural policy); Jennifer Fulton, Chief Executive of Ulster Wildlife (who also helped on water and carbon questions); Lucy Rogers, Director of Conservation Programmes, Avon Wildlife Trust; Lucy Watts, Manager, Vine House Farm; Matt Shardlow, CEO, Buglife; Simon Tonkin, Conservation Manager, Conservation Grade; Ellie Crane, Agricultural Policy Officer (Pesticides and Pollinators) at the RSPB; Dr Alistair Leake, Director of Policy, Game & Wildlife Conservation Trust; and Bill Jordan, Founder, Conservation Grade & Jordan's Cereals.

In writing the chapter entitled 'Have our fish had their chips?', I was helped by: Professor Martin Attrill, Director of the Marine Institute, University of Plymouth; Charles Clover, Founder of the Blue Marine Foundation; Joan Edwards, head of Living Seas, The Royal Society of Wildlife Trusts; John Goodlad, Chairman, Shetland Catch, Shetland Fish Products and the Scottish Pelagic Sustainability Group and advisor to The Prince of Wales's International Sustainability Unit; James Simpson, Communications Officer, Marine Stewardship Council and Tom Hooper and Euan Dunn of the Marine Policy team at the RSPB.

On the role played by freshwater in supporting Britain's health, wealth and security I was grateful for the help of: Dr Charlie Bell, Living Rivers Officer, Hertfordshire & Middlesex Wildlife Trust; Rob Cunningham, Head of Water Policy at the RSPB; Debbie Tann, Chief Executive, Hampshire & Isle of Wight Wildlife Trust; Martin de Retuerto, Area Head of Conservation (Central Rivers & Downs), Hampshire and Isle of Wight Wildlife Trust; Andy Guy, Senior Adviser, Natural England; Prof. Richard Brazier, Associate Professor of Earth Surface Processes, University of Exeter; Harry Barton, Chief Executive, Devon Wildlife Trust; Peter Burgess, Project Development and Policy Planning, Devon Wildlife Trust; Lewis Jones, Future Quality Obligations and R&D Manager, South West Water; Dylan Bright, Head of Sustainability, Upstream Thinking, South West Water; Roy Taylor, Catchment Manager, Northern Ireland Water; Andrew Walker, Catchment Manager, Yorkshire Water; David Harpley, Conservation Manager, Cumbria Wildlife Trust; Rob Stoneman, Chief Executive, Yorkshire Wildlife Trust; Sian Chapman, Corporate Communications Manager, Nestle Waters UK.

During the course of writing *What Nature Does For Britain* my work as an Associate with the economic consultancy Indepen led me to be involved with an investigation into the economic value of actions in catchments to protect and enhance aquatic environments. From this I gained several valuable insights and am grateful to Indepen colleagues John Hargreaves, Ann Bishop, Felicity Furness and Rob Scarrott.

On the subject of flooding, I was assisted by: TV Producer, author and naturalist Stephen Moss, who among other things guided Lucy and me around the Somerset Levels at the height of the 2014 flooding; Clive Faulkner, Trust Manager, Montgomeryshire Wildlife Trust and his colleague Liz Lewis-Reddy, Head of Living Landscapes, Montgomeryshire Wildlife Trust; Roisin Campbell-Palmer, Field Operations Manager, Royal Zoological Society of Scotland; John McAllister, Head of Reserves, East Kent, Kent Wildlife Trust (who gave valuable insights into how beavers help to manage wetlands); John Badley, Senior Sites Manager, Lincolnshire Wash Reserves, RSPB; Dr Rachel Dearden, Hydrogeologist, British Geological Survey, Environmental Science Centre, Gary Grant, Director of the Green Roof Consultancy and Alastair Driver, National Biodiversity Manager at the Environment Agency.

On the capture and storage of carbon in natural systems I was helped by: Dr Nicola Beaumont, Environmental Economist at the Plymouth Marine Laboratory; Darren Hopkins of the British Biochar Foundation; Nick Atkinson, Senior Advisor, Conservation Team, Woodland Trust; Olly Watts, Senior Climate Change Officer, RSPB; Elspeth Ingleby, Chat Moss Project Officer with the Lancashire Wildlife Trust, and Wendy Tobitt at the Berkshire, Buckinghamshire and Oxfordshire Wildlife Trust, who helped explain the work they are doing at Bernwood Forest.

On how best to join up the protection of nature with action to fight climate change, I was assisted by: Mark Campanale, Financial Analyst & Founder and Executive Director, Carbon Tracker; Sophie Bennett, Skills and Employment Policy Officer and Dr Gordon Edge, Director of Policy, both at Renewable UK; Gwyn Willams, Head of Reserves and Protected Areas, Ivan Scrase, Energy and Climate Change specialist, and Mark Williams, Casework Officer, all at the RSPB; Mark Shorrock,

Chief Executive, and Andy Field, Head of Communications, with Tidal Lagoon Power and Dr Jeremy Leggett, Non Executive Chairman, Solarcentury.

The question of how nature supports our health and well-being is an especially fast-expanding area of enquiry and I was fortunate to enjoy assistance from a wide range of experts and practitioners: Dr Jo Barton, Lecturer in Sport and Exercise Science, Dr Rachel Bragg, Senior Research Officer, Carly Wood, Research Officer, John Wooller and Mike Rogerson, all with the University of Essex; Tim Mackrill, Senior Site Manager (Projects), Rutland Water Nature Reserve, Leicestershire and Rutland Wildlife Trust; Andrew Brown, Head of Sustainability, Anglian Water; Dr Rob Lambert, environmental historian at the University of Nottingham; Phil Burfield, Education Policy Officer at the RSPB; Ken Ryan, Senior Communication Officer at MIND Mental Health Charity; Tony Li, Grants Officer, MIND Mental Health Charity; Sue France, CEO, Green Estate Ltd; Nigel Doar, Director of Strategy, The Royal Society of Wildlife Trusts; Vicky Charnock, Arts Coordinator, Alder Hey Hospital; Robin Maynard, Director of Community Programmes at the Avon Wildlife Trust; Gary Mantle, Chief Executive with the Wiltshire Wildlife Trust; Linda Baldwin, Community Engagement and Learning Manager with the Sheffield and Rotherham Wildlife Trust; Simon Phelps, Youth Engagement Officer, Warwickshire Wildlife Trust.

I am also grateful for the assistance and advice of Donal McCarthy, Environmental Economist at the RSPB; Stephen Trotter, England Director, RSWT; Adam Cormack, Head of Communications at RSWT; Graham Tucker at the Institute of European Environmental Policy; and Sophie Churchill, Chief Executive, National Forest.

I am also grateful to my publishers, Profile Books, and in particular to my editor Mark Ellingham, who provided much valuable advice and support, as well as expert editorial input. I am also pleased to have worked with publicist Anna-Marie Fitzgerald, who helped make sure the book received attention. I would also like to thank Tony Lyons, who came up with such a marvellous cover design, and Judy Oliver for a fine index.

Finally, and as ever, I must thank my lovely wife Sue Sparkes and our children Maddie, Nye and Sam for their endless patience with my often all-consuming work.

I've tried to avoid mistakes, but if there are any they are mine alone.

In addition to the wisdom, advice and reflections of all the experts mentioned above, I also relied on a great deal of excellent published material. Interested readers will find access to the sources used to write *What Nature Does For Britain* at my website (**www.tonyjuniper.com**). Click on the book and go to 'read more' and you'll find references organised by chapter.

Index